Ideology and Collapse:
Risks of Climate Science Dismissal

William McPherson

Copyright © 2019

ALL RIGHTS RESERVED.

This book contains material protected under International and Federal Copyright Laws and Treaties. Any unauthorized reprint or use of this material is prohibited. No part of this book may be reproduced or transmitted in any form or by any means, electronic or mechanical, including photocopying, recording, or by any information storage and retrieval system without express written permission from the author.

About the Author

William McPherson is a retired diplomat with experience in global environmental policies. He was assigned to Geneva with responsibilities for reporting on climate research and other UN environmental activities. Following retirement, he continued work on environmental conferences with *Earth Negotiations Bulletin*, the reporting service of the International Institute for Sustainable Development. He is the author of four books on climate change: *Ideology versus Science, Climate (2014), Weather and Ideology (2015), Sabotaging the Planet (2016),* and chapters in *Climate Abandoned (forthcoming)*.

Contents

Collapse 4
Ideology Versus Science 19
Climate, Weather and Ideology 33
Sabotaging the Planet 64
Climate Abandoned 98
Ideology and Collapse 112
Appendix 161

Collapse

Unless humanity is suicidal (which, granted, is a possibility), we will solve the problem of climate change. Yes, the problem is enormous, but we have both the knowledge and the resources to do this and require only the will. E.O. Wilson[1]

In Jared Diamond's book *Collapse*, the author explores how neglect of environmental factors led to the downfall of societies in places such as Greenland and Easter Island. In a nutshell, it was the hubris of inhabitants of these ecologically fragile areas that caused their collapse, a hubris based on their beliefs that natural forces would not apply to them. I call this collapse ideology. Those who promote policies leading to collapse are collapse ideologues. "Collapse" is intended as a double *entendre*: as consequences of climate change become more obvious, the ideology itself may collapse – because it is unsustainable in the face of change. It will of necessity be abandoned, but that may come too late.

For background on climate, we can start with trends toward collapse in the atmosphere, geosphere and biosphere, as demonstrated by the following data:[2]

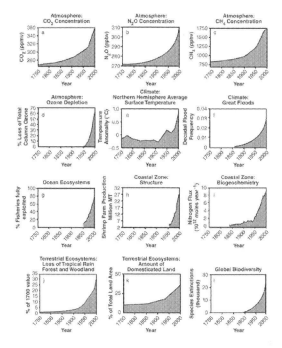

Each of these graphs indicates exponential, ever-increasing growth in use of earth's resources, trends that lead to collapse. The first six graphs illustrate the causes and effects of atmospheric changes. CO_2 (carbon dioxide), N_2O (Nitrous Oxide) and CH_4 (methane) are greenhouse gases, and other graphs show ozone depletion, temperature and flooding. The next six graphs show the effects humans are having on the biosphere through fisheries and agriculture, plus the general effects of all these changes on biodiversity. In other words, the trends indicate that we are using resources, including the carbon-absorption capacity of the atmosphere, at an unsustainable rate that will lead to overshoot and collapse.

There is no better description of collapse than the speech by David Attenborough to the climate meeting in Katowice, Poland in late 2018. He said, "If we don't take action, the collapse of our civilisations and the extinction of much of the natural world is on the horizon."[3]

What ultimately causes collapse? We know that in the cases of Easter Island and Greenland there were cultural factors that led inhabitants to believe that they were not subject to environmental forces. Easter Island residents clearcut the island for logs to transport the large heavy memorials to the deceased. Greenland residents from Europe ignored the cooling of the climate after the end of the Medieval warm period; it may be that their religious beliefs gave them hope and solace when things got worse. Up to the end, they were still practicing European-style agricultural methods and attending churches. They could have survived if they had adopted the food habits of the Inuit people who also lived and thrived on Greenland, even in its colder periods.

It is clear that religious beliefs, particularly beliefs about death and afterlife, have influenced societies such as Easter

Island and Greenland. The monuments in Easter Island were images of ancestors, beatified to symbolize their immortality. The practices of the Greenland Vikings were based on their church teachings, including the quest for immortality as a solace for their declining living situations.

Religion is not an unmitigated source of collapse ideology, however. We will be looking at both sides of religion, the tendency toward anthropocentrism as well as the moral motivation to relate to nature on its own terms. The former is a more dangerous form of religion, which leads to tendencies to abandon nature, while the latter is more hopeful when it comes to confronting climate change.

Tracing the effect of ideology on societal failures is difficult, because many other forces impinge on behavior. I have attempted to do this analysis on a variety of social trends, ranging from social movements[4] to war[5] and various aspects of climate science denial.[6]

In *Ideology and War,* I explore how Nazi Germany's and Imperial Japan's belief in their own superiority led to their downfall. Germany's belief in *Übermensch* (racial superiority) and Japan's belief in *Bushido* (the way of the warrior) led to imperial overreach and eventual collapse.

They are examples of collapse ideology at the national level.

Now we are faced with global hubris and neglect of environmental forces, leading to collective suicide from climate change. In *Ideology versus Science,* I document the confrontation of belief versus fact that has led us to avoid difficult choices affecting continued growth in the use of fossil fuels and the consequences for the natural ecosystem. Most often this takes the form of an economic/political belief in growth, a kind of faith in unlimited natural resources available to sustain continuous increases in production and consumption.

Growth

Assumptions of growth affect more than political ideology. Economists have built it into their models. The danger lies in the omission of outside factors: "Climate change is having a real impact, not just on the environment but on the economy too. …But almost no mainstream economic forecasting model takes that into account, in an omission that some economists say could affect the accuracy of economic predictions going forward."[7] While climate scientists are sometimes accused of using inaccurate models, the denial of climate change consequences also affects economic models. There are major risks to ignoring climate effects in economic analysis. This is particularly true of economic analyses of energy.

Often the approach takes the form of promoting the advantages of using fossil fuels, as in this paragraph from Bastardi:

> I could not possibly have had the life I've had or do what I do (this applies to all of us) without fossil fuels. Which is one of the things that bugs me so much about the whole climate alarmist movement. Here we are, all nice and warm with our marvelous technology, and the same people and industries that help so much to make it that way are now demonized and destroyed. I believe the word to describe someone that does that is "ingrate."[8]

I guess we are *ingrates* if we dare to suggest that fossil fuels have a downside as well as all the benefits attributed to them. Perhaps we are guilty of suggesting that there are alternatives to fossil-fueled energy systems. We are even more guilty if we might suggest that the reliance on fossil fuels would lead to collapse.

Some would suggest that those urging climate action are promoting a return to primitive society, "living in caves" and giving up all of the comforts of modern society. This is simple-minded, as there are many ways to attain a comfortable lifestyle that do not require unlimited growth in fossil fuel use. Nevertheless, the inertia of belief continues to blind us to the possibilities of change without deprivation. As Frank Incropera puts it, "A world in which ever increasing numbers of people are consuming ever more is not sustainable. That is not to say humanity must move to a state of deprivation. It is possible to lead comfortable and fulfilling lives while moderating our tendencies for self-indulgence and overconsumption. Living big is not a prerequisite to living well."[9]

Interestingly, the fossil fuel industry itself is beginning to wake up to the idea that growth in the use of fossil fuels cannot continue indefinitely. Senator Sheldon Whitehouse (D-RI) has noted: "You now have a situation in which the CEOs of Shell and Exxon and the other big oil companies have to publicly admit that their products are causing climate change."[10] Of course, as Senator Whitehouse noted, these companies continue to explore for additional oil fields even as their current reserves are five times more than can be safely burned. This is a prime example of collapse ideology.

Ideology can have a peculiar effect on people's lives. While clinging to their beliefs, consumers continue self-destructive behavior, even behavior they know will lead to their eventual demise. This is the most significant effect of climate science dismissal, a collapse ideology that is expressed not only anti-scientific beliefs, but in the day-to-day actions of people using fossil fuels unwisely. In *Climate, Weather and Ideology,* I discuss how the self-destructive behavior of the U.S. public embedded in the political, economic and cultural milieu.

Extreme Weather

Among the more egregious aspects of collapse ideology is the avoidance of climate analysis in reporting extreme weather. "From September 9 through September 16, [2018], only 7.5 percent of pieces in the top 50 newspapers that discussed Hurricane Florence mentioned climate change. Only 4.3 percent of major broadcast segments on the storm mentioned it."[11] Yet scientists estimate that rainfall was 50% higher because of climate change, and flooding was the cause of most of Florence's damage. If people cannot be informed of the link between damage and climate change, how will they be motivated to act on it?

Of course, humanity is not unaware of the crisis and many are willing to take actions to avoid collective suicide. In *Sabotaging the Planet,* I describe the efforts that led to the Paris Agreement in December 2015. Unfortunately, the pledges to reduce emissions were insufficient to meet the two main Paris goals, (1) keeping temperature increases "well below" 2 degrees Centigrade with aspirations for a 1.5 degree limit, and (2) net zero emissions by the "second half of the century," ideally 2050.[12] Net zero emissions require that residual positive emissions be equally offset by negative emissions, i.e. sequestration of carbon through, for example, reforestation. That is a tall order, but an easily measurable goal. Politicians and civic leaders will have to do the math and get on a pathway that *reduces* emissions by at least three percent a year. Currently we are *increasing* emissions by more than three percent a year.

The UN Environment Program (UNEP) reviews emissions relative to the commitments under the Paris Agreement. UNEP found the following in 2018: "Global greenhouse gas emissions show no signs of peaking. Global CO2 emissions from energy and industry increased in 2017... Total annual greenhouse gases emissions, including from land-use change, reached a record high of 53.5 GtCO2e [gigatons of carbon dioxide equivalent] in 2017, an increase of 0.7 GtCO2e compared with 2016. In contrast, global GHG emissions in 2030 need to be approximately 25 percent and 55 percent lower than in 2017 to put the world on a least-cost pathway to limiting global warming to 2C and 1.5C respectively."[13] We are not on track to meet the Paris commitments, and politicians seem to be neglecting their countries' commitments or ignoring them altogether.

Even worse, sabotaging commitments are politicians such as Donald Trump who not only abandoned the Paris Agreement but also encouraged emissions increases with his energy policies. Many other politicians in Congress and some state governments supported him. They exhibit the symptoms described in the forthcoming book *Climate Abandoned.*

Not all politicians are involved in collapse ideology. A Democratic candidate for president, John Delaney, has recognized the problem: "Obviously, a transformative government response to climate is clearly needed based on what we're seeing from science. We have about 10 years left to really be doing something, but it's hard to do because the Republican party is largely in denial on this."[14] While it is true that many Republicans are dismissing climate science, theirs will be a decreasingly credible position as climate change consequences become obvious. Most Republicans, in fact, agree with climate science although they may not agree on measures to address it.[15]

Pollution Paradox

One explanation for the persistence of climate inaction is the Pollution Paradox, described by George Monbiot as follows: "The dirtiest companies must spend the most on politics if they are not to be regulated out of existence, so politics comes to be dominated by the dirtiest companies. The paradox applies across the board. Banks designing dodgy financial instruments; pharmaceutical companies selling outdated drugs at inflated prices; gambling companies seeking to stifle controls; food companies selling obesogenic junk: all have an enhanced incentive to buy political space, as all, in a functioning democracy, would find themselves under pressure. Their spending crowds out less damaging interests, and captures the system."[16] *Crowding out less damaging interests* is a key concept for understanding why politicians are reluctant to pursue climate action. A majority of the U.S. population supports climate action but it is a "less damaging interest" than the concentrated power of the fossil fuel interests. Politicians view fossil fuel companies as more "damaging" in their use of political influence than climate activists. Once climate consequences become more damaging, however, that may change. Will it be too late? That is a question we will try to answer.

The Pollution Paradox is also a moral issue in economics. One sometimes hears the Milton Friedman argument, that corporations have only one obligation: "The social responsibility of business is to increase its profits."[17] This is not an economic argument; it is a moral argument. There is no reason that economies have to be run solely on the basis of maximizing profits. Corporations are, after all, organizations of people that are endowed with certain rights. Organizations with these legal powers do have moral obligations other than making profits, even if one speaks only of survival. As we will see, survival is the most problematic issue of current history.

As a victim of burglaries, I know this "here and now, get rich quick" mentality pervades not only the corporate world, but also the societies in which they operate. Few people are willing to consider the long-term risks of their behavior when they see an immediate opportunity to get something "free." Someday we will all be victims of this mental criminality that pervades collapse ideology.

Political Resistance

From early in the history of climate science, scientists recognized that when they demonstrated the effects of carbon emissions, they would not find a receptive audience among politicians. "'Do we have a problem?' asked Anthony Scoville, a congressional science consultant [at a 1980 conference]. 'We do, but it is not the atmospheric problem. It is the political problem.' He doubted that any scientific report, no matter how ominous its predictions, would persuade politicians to act."[18]

An example of politicians supporting collapse ideology is found in the statement on the website of Senator Ted Cruz:

> U.S. Sens. Ted Cruz (R-Texas), Rand Paul (R-Ky.), James Lankford (R-Okla.), and Jim Inhofe (R-Okla.) this week requested National Science Foundation (NSF) Inspector General Allison Lerner investigate the NSF's grant-making process, relaying their concern that the NSF has "issued several grants which seek to influence political and social debate rather than conduct scientific research" in contradiction of federal law and the agency's mission. Among the examples, the senators cited the NSF providing over four million dollars to a climate-change coalition to turn television meteorologists into *climate change evangelists* (with almost three million coming after initial research revealed a significant lack of consensus on climate change among meteorologists), as well as a grant to increase the engineering industry's activism on social justice issues.[19]

And what did NSF do to make television meteorologists "climate change evangelists?" *It proposed that they report the facts.* That seems to bother politicians.

Politicians' views on climate are supported by ideologues whose interest in climate change seems to be to make a career of denying it. Cornwall Alliance insists that the cold and snowy weather of the winter of 2017-18 shows that temperatures are not increasing. Nevertheless, the data shows otherwise: on a worldwide basis, temperatures increased during that winter. A regional variation from the global average does not prove that climate science is wrong.

Other ideologues described in the book *Ideology versus Science* continue to dismiss aspects of climate science. For example, Fred Singer, who is associated with the Heartland Institute, has claimed that sea level rise is not due to global warming: "sea-level rise does not depend on the use of fossil fuels. The evidence should allay fear that the release of additional CO_2 will increase sea-level rise."[20] This misstatement resembles a law passed by the North Carolina legislature, when it decreed that shoreline planning should not take into account any increase in the pace of sea level rise due to climate change. One might call it the King Canute syndrome. (By the way, King Canute was not really intending to hold back the tide. He would try it in front of his courtiers, just to

17

demonstrate it was futile and that nature is in charge. We could learn from this.)

In the chapters that follow I intend to wrestle with these issues and try to find some explanations. I have been researching this area for ten years and have published four books on the subject of climate denial. I will draw from these books and add updates where appropriate.

Ideology Versus Science

It's the new form of denialism: First it was that climate change isn't happening, then it was that humans weren't contributing, then it was pointing out the other sources – what about the sun, what about volcanoes? – and now it's going with the most moderate projections and playing up the uncertainty of the science. Naomi Ages.[21]

I must say at the outset that I find it difficult to understand how people can vigorously and blatantly dismiss climate science. I think it is less mystifying how people can passively deny, or ignore, the more serious and threatening aspects of science. They do not want to hear about something that upsets their view of the world and makes them think about changing habits or patterns of consumption. "Leave me alone" is the common form of this view. Camilo Mora of the University of Hawaii at Manoa explained this: "We as humans don't feel the pain of people who are far away or far into the future. We normally care about people who are close to us or that are impacting us, or things that will happen tomorrow. We can deal with these things later, we have more pressing problems now."[22]

But the deliberate ginning up of some blatant pseudoscience explanations to brush away the findings of climate science is a different matter. Of course, it is usually done at the behest of funding organizations, such as the Koch Brothers' "Americans for Prosperity," the Heartland Institute and the right wing think tanks, but money alone does not explain the ideological tendencies of collapse ideologues. *They really want to dismiss climate science,* a strange phenomenon in a world where respect for most scientists, and action on the basis of understanding of real-world information, tends to guide most people's behavior. Unfortunately, in the case of climate change, science credibility is not respected. Many people cling to false beliefs or "alternative facts" that are increasingly threatened by reality. Their world view will weaken as the real world intrudes on collapse ideology, when storms, floods and wildfires call into question the beliefs of the ideologues.

Along with outright denial there is the phenomenon of doubt. "Doubt is our product" was the motto of organizations that tried to stop anti-smoking campaigns on behalf of the tobacco companies, and it is now the favorite strategy of collapse ideologues. By emphasizing uncertainties in the science, they prey upon the desires of listeners to avoid dealing with climate change.

There are some sophisticated analyses of climate science that do find faults in the research and enumerate the uncertainties in the science.[23] But to argue that some faults and uncertainties undermine all the findings of the scientists is a collapse ideology completely divorced from reality. Any such problems with climate science do not mean it is a conspiracy to fool people into doing foolish things. It is a fundamentally serious warning about the effects of burning fossil fuels with plenty of evidence about what has already happened, and plenty of validity in its projections of what will happen.

An example of the kind of misreading of science that persists is in a recent book by Marc Morano, *The Politically Incorrect Guide to Climate Change:* "…in fact, the claimed 'record heat' is within the margin of error between so-called hottest years – a fancy way of saying the temperature standstill is continuing."[24] Morano is one of a group of ideologues that assert that temperatures have stabilized despite evidence to the contrary:[25]

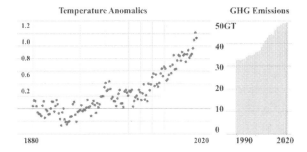

I marvel at the ability of some authors to hold contradictory positions about climate. Unfortunately for Morano, temperature is increasing at an accelerating rate.

Collapse ideology has another insidious effect on climate science: it discourages the research necessary to understand climate change, particularly government science. Michael White noted, "This federal investment in climate change research is working as intended: We now know more than ever about the causes and consequences of climate change, and the possible ways to avoid or adapt to it. But the work of our science agencies has provoked a furious backlash, one that goes beyond mere debate over the risks we face and the solutions we should pursue. This backlash of climate change denial is killing our ability to act, by attacking the very research institutions that we established to help us solve our problems."[26]

Collapse ideology has its effects on government support for climate research in numerous ways. Most of the Trump Administration cabinet departments – EPA, Treasury, Interior, Agriculture, National Institute for Occupational Safety and Health, and others[27] – have now removed climate change pages and references to climate change from government web sites. While this in itself does not mean that the research has stopped, it does mean that public perception of the issue is dampened and its priority in policymaking is severely diminished.

Ironically, many executive agencies continue their work on climate and report that current trends in the use of fossil fuels will lead to collapse. A legally-required report, the National Climate Assessment (NCA), was issued in November 2018. The National Highway Transportation Agency, for example, has reported that temperatures will increase 4C by 2100 if current trends continue.[28] That increase will make any chance of continuing with "business as usual" in the transportation sector moot. When this information at a working level filters up to political leadership, collapse ideology is likely to suppress it at the higher levels. That is exactly what happened with the NCA; President Trump dismissed it by declaring "I don't believe in it." He also claimed that his "very high level of intelligence" enables him to reject the NCA, the IPCC report (see Appendix) and a climate report by the World Meteorological Organization.[29]

Where is the EPA?

EPA is administered by Andrew Wheeler, a former lobbyist for Murray Energy. The EPA is relaxing rules on emissions under the Trump Administration to reduce utility costs, a clear example of collapse ideology. Wheeler has justified rolling back some Obama-era regulations by claiming that the Obama administration's calculation of health benefits was "suspect." He said that the Trump administration's rollback of vehicle emissions rules and the Clean Power Plan would not affect public health protections. "Neither one of those is health-based standards per se. We have our separate health-based standards."[30] It is a favorite tactic of collapse ideologues to promote standards separate from evidence-based science.

Some might say that EPA appointees do not know what they are doing, but it appears that they know exactly what they are doing. They are implementing collapse ideology deliberately and systematically. For example, when the EPA proposed changing the Obama-era methane rule, they promoted the release of methane at a much higher rate.[31] Cecilia Munoz, a former

Obama policy director said, "...they appear to have brought into the administration ideologues who don't know a lot about policymaking, but in climate change and energy, they appear to have brought in people who know exactly what they're doing, and know exactly where the levers are".[32] What they are doing is deregulating a predatory fossil fuel industry bent, in the most rapid way possible, on depleting the capacity of the atmosphere to absorb carbon. "Let's get going quickly, before people discover how much damage our industry creates," seems to be the working philosophy.

By ignoring the health consequences of deregulation, the Trump Administration officials are pandering to a business crowd. While many business people are conservatives, this group does not really follow conservative principles. A group of Republicans including James Baker, George Schulz and Henry Paulsen have endorsed a carbon tax.[33] There is controversy about the regressive nature of a carbon tax, but many economists favor it. Incropera contends that it would be the most effective policy.[34] If it is coupled with a dividend to alleviating the regressivity, it would be a more equitable approach. It is not politically viable, however; the opposition from Congress is discussed below.

Nor can we blame business in general for collapse ideology. Some businesses, such as insurance and agribusiness, recognize the risks and plan for climate change. General Mills CEO Ken Powell said, "We think that human-caused greenhouse gas causes climate change and climate volatility, and that's going to stress the agricultural supply chain, which is very important to us. Obviously, we depend on that for our business, and we all depend on that for the food we eat." Unilever, Mars, and Nestlé have set targets similar to those of General Mills: reduce emissions by 28 percent by 2025.[35]

Decline of Denial?

In the years since I published *Ideology Versus Science,* climate denial ideology has thrived rather than withered as expected. Of the movements that I analyzed in that book, only the Tea Party has faded, and one could argue that they have gone "establishment" in that a majority of Republicans in Congress and executive offices now seem to follow in their footsteps. For example, President Trump and Tea Party darling Energy Secretary Rick Perry continue to promote fossil fuels. They demonstrated this in a rule requiring generating stations to keep a 60-day supply of fuel on site. This seems to promote coal – how would a gas-fired plant keep a 60-day supply on site – but it really ignores the advantages of renewable energies. Wind and solar fuel supplies are unlimited![36]

Rick Perry was one of the strongest collapse ideologues among Republican candidates when he ran for president in 2016. Perry said climate science is "all one contrived, phony mess that is falling apart under its own weight." As governor of Texas, Perry regarded consequences of climate change, such as droughts, as "acts of God" best addressed through prayer. Candidate Perry also prayed for a change in national policies. In his book *Fed Up*, Perry used an argument that often appears in denial ideology – that regulation of carbon emissions is not authorized by legislation and would have "devastating" economic effects. Perry also used denial ideology to attack Democrats. He asserted "science gets hijacked by the political Left" and "may not stand the test of time."[37] Apparently Perry's previous experience and beliefs prepared him well to serve as Trump's Energy Secretary.

Other groups such as Heartland Institute, and the Koch-funded Americans for Prosperity and American (AFP) and American Legislative Exchange Council (ALEC), have maintained and even increased their membership and influence. ALEC and AFP have been active in opposing renewable energy and mass transit. AFP, for example, has been instrumental in killing a light rail project in Nashville and has opposed similar projects in several other cities around the nation.[38] These kinds of actions are indicative of a collapse ideology, where preventing necessary changes in the infrastructure now will lead to major problems in the future when climate consequences become unavoidable. We will be forced to do climate crisis management by the AFP and ALEC actions.

David and Charles Koch have funded all these efforts. Even though they were unhappy that Donald Trump was elected president, they were happy to see the EPA and other government agencies pursue their deregulation agenda.[39] The Koch brothers have long dismissed climate science and worked to impose on us their one-sided views of fossil fuel growth, promoting collapse ideology.

One reason groups such as AFP are successful is that they are influencing teachers and, through them, students, with claims that the science is uncertain.

> A 2016 national survey of 1,500 science teachers found that 31 percent told their students that global warming is still up for debate. One in 10 told students that climate change is a natural phenomenon. Five percent simply didn't mention the subject. And a 2015 Pew survey

found that while young adults aged 18 to 29 agree that humans are causing climate change more than their parents and grandparents do, 37 percent believe there is either no solid evidence for it, or that it is natural.[40]

What this means in terms of climate action and policy is ominous. Delay of necessary changes in energy infrastructure and transportation are serious setbacks, and effective action will be more difficult and costly as a result. Congress and state legislatures are hemming and hawing, even in blue states such as Washington where the legislature failed to pass a carbon tax in either 2016 or 2018. The Washington State initiative for a carbon fee was defeated when oil companies spent more than $30 million to defeat it.[41] It is clear that collapse ideology, promoted by oil companies, has had a major influence on elections.

A good example of the election of a candidate that dismisses climate science is the success of Rick Scott, former governor of Florida, in defeating Bill Nelson in the 2018 campaign for U.S. Senate. Scott, as governor, had prohibited mention of climate change in state documents, even though his state is vulnerable to sea level rise. Miami often floods from sea water seeping into the storm sewer system and covering streets with salt water, even when there is no rain.

The effects of climate change will not be delayed by dismissal of climate science; the only thing being delayed is effective climate action.

Delay is an insidious effect of collapse ideology. Johansen describes this as follows: "Global-warming deniers kept change at bay, it may be noted, by appealing to most people's fear that change might erode their comfort and employment security – all of which were wedded psychologically to the massive burning of fossil fuels. A necessary change in our energy base, they may conclude, may have been stalled beyond the point where climate change required attention, comprehension and action."[42] Stalling this *necessary change* is a result of collapse ideology, that keeps societies from avoiding climate consequences until it is too late. Feedbacks and other forces inherent in the climate system can make late action ineffective, when earlier action could have prevented climate consequences at much lower cost.

Market Fetishism

One kind of political philosophy that looms large in collapse ideology is "markets know best." The fetishism of markets tends to blind people to consequences of action that harm other people. For example, the Koch-funded Americans for Prosperity have fought renewable energy and mass transit on the basis that they require government actions rather than private sector market decisions. They stir up opposition to funding infrastructure that would reduce the need for fossil fuels. This is a political philosophy that would, indirectly and unintentionally, lead to collapse in the long run.

It is bizarre to read books that describe a world totally different from the one in which we live. It is a world in which temperature increases have stopped: Tim Ball claims that "the decline in temperature since 1998 occurred while human production increased more than in the period before 1998."[43] This is a false statement, as we shall see. It is used to portray a world where there is no reason to discuss human contributions to global warming because there are none. It is a world where heat waves, wildfires and floods are just part of a natural cycle, and we just think they are getting worse. It is a world where scientists, who are trying to alert us to consequences of climate change, have nefarious motives such as greed.

For authors of these books, there seems to be a disconnect between reality and the fantasy world in which they want to live. Of course, this disconnect has been discussed extensively in studies of cognitive dissonance as it applies to climate science dismissal.[44] Denial book authors do not want to live in today's world; they seem to construct a nostalgic version of the past where one does not have to face climate consequences. It makes them feel better to align the world with their beliefs than to align their beliefs with the world.

Theirs is not a world in which I live. When I read these books, I have a queasy feeling about how humans can distort reality to fit their preconceptions. I know there is a tendency toward confirmation bias in controversial areas, but climate science should not be controversial. It is only made so by the special interests that want to stop our understanding of the climate system.

Climate, Weather and Ideology

> *Climate normality seems to be built into our world. We experience the climate as part of "eternal nature." Actually, climate change has been ... brought about by two hundred fifty years of industrialization, with dangerous spikes in recent decades. But the process has been largely silent.* Robert Lifton.[45]

As Lifton indicates, we have been taken by surprise by something that has been brought about by two hundred fifty years of industrialization and only recently have we noticed "dangerous spikes" in the weather. We maintain a belief in "climate normality," and when confronted with abnormality we tend toward willful ignorance, a defense mechanism that operates at both the personal and political level. But nature does not cooperate. "The brutal weather has been supercharged by human-induced climate change, scientists say. Climate models for three decades have predicted exactly what the world is seeing this summer."[46]

Interestingly, the general public is seeing what some political leaders do not see. "…there's lots of evidence that contemporary weather is a contributing factor to belief in climate change," according to the director of the Muhlenberg College Institute of Public Opinion, Chris Borick. "The talking points have turned more to the cost to mitigate climate change rather than deny its existence."[47] When Hurricanes such as Harvey and Florence made landfall in 2017 and 2018, they dumped much more rain on communities and caused worse floods than previous hurricanes of the same strength. This got people's attention because many more were flooded out than before. The general public is beginning to understand that climate change is here and it is dangerous, but their political representatives are still not with them.

While the world burns, political leaders fiddle. The situation has not improved much outside of the realm of political inaction. Dismissal of climate science persists in large segments of the political class – governors of Florida, Maine, and others, many senators and members of Congress and state legislators. In the U.S., politicians are increasingly out of touch with the public. At a global level, the U.S. is increasingly out of touch with other countries where climate is having severe impacts. A prime example is Steven Milloy, a member of Trump's EPA transmission team, who dismissed the findings of the government's own climate researchers: "We don't care. In our view, this is made-up hysteria anyway."[48] *"Made-up hysteria"* hardly describes the work of hundreds of government and academic scientists who contributed to the report, *Fourth National Climate Assessment*.[49] A more bland statement by President Trump: "I don't believe it." Apparently he does believe in fossil fuels, however. His press secretary, Sarah Huckabee-Sanders, said the assessment was "not based on facts." Perhaps she relies on the "alternative facts" found in collapse ideology. Interior Secretary Ryan Zinke elaborated that the Trump Administration disowns the work of scientists, including its own government officials, because they used "extreme scenarios" in predicting climate problems. EPA Administrator Andrew Wheeler went even further, saying "I don't know for a fact – I wouldn't be surprised if the Obama administration told the report's authors to take at a look at the worst case scenario for this report."[50] One of the authors, Katherine Hayhoe, responded that "I wrote the climate scenarios chapter myself so I can confirm it considers ALL scenarios, from those where we go carbon negative before end of century to those where carbon emissions continue to rise. What WH says is demonstrably false."[51]

Climate Consequences

Examples of extreme heat have multiplied in recent years, particularly in countries with large numbers of poor people: "… as many of South Asia's already-scorching cities get even hotter, scientists and economists are warning of a quieter, more far-reaching danger: Extreme heat is devastating the health and livelihoods of tens of millions more. If global greenhouse gas emissions continue at their current pace, they say, heat and humidity levels could become unbearable, especially for the poor."[52]

South Asia is not the only part of the continent where extreme heat has become more common. "'This heat is a threat to life,' the Japanese Meteorological Agency warned. 'We recognize it as a natural disaster.'…While the heat wave certainly has natural roots, it's impossible to talk about what's happening in Japan without including climate change. The heat that's built up on our planet has made heat waves more common and more intense."[53]

People are beginning to recognize that the weather is not what it used to be. Even in advanced countries such as the U.S. and Sweden, high temperatures and wildfires make the effects of climate change obvious. "In the past, scientists have been reluctant to cite climate change as a major factor in California's worsening wildfires…But the connection between rising temperatures in California and tinder-dry vegetation is becoming impossible to ignore, according to experts who study climate and wildfires."[54] California fire chiefs have said that the wildfire season is now year around; in Sweden, a wildfire was burning north of the Arctic Circle. While recognizing changes, people do not always understand the source of those changes. Although more than half of the population now agrees there is climate change, less than half seems to think that is caused by humans. People seem to cling to false beliefs as documented in my book *Climate, Weather and Ideology*,[55] fueled by dangerous ideas that continue to appear on bookshelves.[56] Nearly every denial author tends to exaggerate and distort reality beyond even the more extreme versions of denial I described in my book.

Why do people persist in dismissing climate science? A glib answer would say that people want to continue on current levels of consumption that depend on fossil-fueled production. But belief in fossil-fueled growth, even at an unconscious level, pervades the culture. One author has suggested that the fossil-fuel energy system is a kind of religion:

> …the theme of steam-as-omnipotence grew out of and fed into the emerging realities of the fossil economy, constituted precisely by the protean nature of the rotative engine. …Time and again, the engine was described as being alive, a subject acting and performing and doing all sorts of things ex *proprio vigore*, even an organism with all the hallmarks of metabolism and vitality.[57]

The author, Andreas Malm, calls this "fetishism," a tendency to objectify and sanctify energy as the driving force of prosperity. He is describing the origins of the fossil-fuel economy in Great Britain at the beginning of the industrial revolution. Unfortunately for subsequent human history, the sanctity of fossil-fueled energy has become entrenched in the world economy. It becomes nearly impossible to detach human welfare from the idea of cheap energy – cheap only because it is priced well below its real cost.

Energy Fetishism

Energy fetishism plays a strange role in the morality of climate change and the place of fossil fuels in this morality. There are some, notably Bjorn Lomborg, who argue that while climate change poses a risk, world poverty is a higher priority. This argument can be twisted into claiming that solving poverty through energy is a higher morality. Lifton describes this spin:

> As we have seen in the case of ExxonMobil, corporate leaders were faced with scientific findings that questioned the overall morality of their companies' activities, which were revealed to contribute to a staggering threat to human beings in general. Such corporations either had to cease these activities or else claim human value for them, such as that of contributing to a better life for millions of people in developing countries, in order to blunt the scientific truths. Here we may speak of an ultimate expression of stranded ethics – a corporation's assertive justification of activities that threaten the human future.[58]

This argument about poverty has been adopted by the Trump Administration. In IPCC reports about keeping warming below 1.5C, the U.S. delegation rebutted the general findings with assertions that global poverty has declined in the recent decades as use of fossil fuel has increased in the developing world. (See Appendix for the IPCC report.) The U.S. criticized the IPCC with a comment: "The report and SPM [Summary for Policy Makers] do not present a balanced assessment of the economic, social and development costs associated with the trade-offs of pursuing actions consistent with limiting global warming to 1.5 C."[59]

During the meeting of the Conference of Parties (COP) of the UN Framework Convention on Climate Change (UNFCC) in 2018, the U.S. State Department issued the following statement:

> The United States was willing to note the report and express appreciation to the scientists who developed it, but not to welcome it, as that would denote endorsement of the report. As we have made clear in the IPCC and other bodies, the United States has not endorsed the findings of the report.[60]

Failing to endorse the report puts the U.S. on the wrong side of history and fails the moral challenge of climate change. The IPCC report clearly lays out the moral culpability of failure to address the climate crisis. (See Appendix)

In these statements, the U.S. contends that other priorities might outrank climate action, an argument consistent with collapse ideology. It has been termed "moral corruption" by scholars working on the ethics of climate change.[61] Without climate action, no other efforts will succeed. Poverty is exacerbated by climate consequences such as changes in precipitation that impoverish farmers in the Middle East, Africa and South Asia.[62] To argue that climate action can be delayed or even denied is a spurious dismissal of the most significant threat facing humanity, one which will render all other problems moot.

Of course, one could not say that the motives of the U.S. for promoting fossil fuels are altruistic. In addition to the State Department statement quoted above, an even more blatant statement was made by President Trump's international energy and climate adviser, Wells Griffith, at the 2018 COP: "The United States has an abundance of natural resources and is not going to keep them in the ground. We strongly believe that no country should have to sacrifice their economic prosperity or energy security in pursuit of environmental sustainability."[63] *In pursuit of environmental sustainability* should be ranked miles above fossil fuel use; in fact, using fossil fuels is a sure sign of environmental disaster which will make "economic prosperity or energy security" impossible.

One of the more egregious versions of these arguments is that climate science somehow distracts us from more important problems. Bastardi argues that concern about climate is a way of avoiding concern about poverty and urban problems.

> …the problems that are occurring now are far more important. In fact, I think one of the reasons people love to obsess over the climate issue is because they don't have to deal with the hard problems – like people starving in our streets, or the decay of our cities, or our culture. So they go after a problem they cannot be judged for since it's so far down the road. What's wild is that many of the policies being advocated by way of the climate agenda worsen the

problems we have now simply by deflecting money and attention from them.[64]

Accusing climate activists of neglecting *"people starving in our streets, or the decay of our cities,"* etc., is a particularly malicious way of dealing with climate change. First, deny that there is a problem, and then accuse those who want to deal with the problem of avoiding more difficult problems. In fact, dealing with climate change now is likely to avoid more severe problems of "people starving in our streets, or the decay of our cities," when the consequences of climate change begin to affect food supply and the livability of cities. Heat waves reduce the productivity of agriculture, regardless of how much additional carbon dioxide is available to plants. Germany's grain production was reduced by 20 percent during the hot summer of 2018.[65] Cities are "urban heat islands," with more surfaces that retain heat and cause problems from heat waves, than other areas. Climate change will indeed lead to "people starving in our streets, or the decay of our cities."

Energy fetishism plays a role in environmental injustice in some strange ways. The generation of electricity, for example, can have a profound effect on some groups such as the Navaho in the Southwest. Not only are they adversely affected by the pollution from coal-fired power, but they also have suffered from uranium mining for nuclear power plants.[66] Even though they are impacted by power generation, many Navaho households do not have electricity. Other minorities are also impacted by the pollution from power generation. Communities of color are often near generating plants and suffer disproportionately from their pollution.

Environmental Justice and Climate Change

Rev. William J. Barber II, a Baptist minister, is fighting for environmental justice in Greensboro, NC. A coal-fired power plant is polluting the waters of a lower-income colored community with coal ash, disposed in unlined waste ponds that contaminate wells. He was joined in late August 2018 by Al Gore, fighting coal power generation as part of his fight against climate change. Gore linked the issues: "Both are necessary byproducts of our addiction to fossil fuels." Rev. Barber put the issue succinctly: "Jesus said love your neighbor. I don't care how many times you tell me you love me, if you put coal ash in my water you don't love me. Because if there was nothing wrong with the coal ash, then put it in the wealthy communities."

Barber's message is a clear example of environmental justice and its connection to politics. Barber said, "This is the real question, not if Democrats are going to get elected, not if Republicans are going to get elected, but if America is going to be America, she's going to have to address systemic racism, systemic poverty, ecological devastation, the war economy and militarism, and our false moral narrative of religious nationalism." He railed against the "*false moral narrative*" that is used to rationalize these problems. It is a narrative inherent in collapse ideology, which finds excuses to continue self-destructive behavior.

Would residents of Greensboro be able to rely on the Environmental Protection Agency (EPA) or the Duke Energy, the operator of the plant, to reduce pollution? The Trump EPA has changed regulations on coal ash to make it easier for utilities to avoid costly storage. Duke Energy has been sued for dumping coal ash into surface water and connected groundwater. It doesn't seem that either EPA or Duke will want to address the problem.[67]

Another issue of environmental injustice arises from heat waves. Data indicate the injustice of climate change: "In the US, immigrant workers are three times more likely to die from heat exposure than American citizens. In India, where 24 cities are expected to reach average summertime highs of at least 35C (95F) by 2050, it is the slum dwellers who are most vulnerable. And as the global risk of prolonged exposure to deadly heat steadily rises, so do the associated risks of human catastrophe." Research shows a trend toward increasingly dangerous heat waves. "Last year, Hawaiian researchers projected that the share of the world's population exposed to deadly heat for at least 20 days a year will increase from 30% now to 74% by 2100 if greenhouse gas emissions are allowed to grow."[68] Exposure to deadly heat is certainly a moral issue that would seem to draw the attention of religious leaders around the world. Not all religions are attuned to the issues of climate change, however.

Religion and Climate Change

Religion plays a strange role in attitudes toward climate change. There is a kind of religious fatalism for people who see environmental destruction all around them but simply want to dismiss any human cause by invoking a divine cause. An example is a Louisiana bayou resident quoted in the book *Strangers in Their Own Land*:

> "The earth will burn with fervent heat." Fire purifies, so the planet will be purified 1,000 years from now, and until then, the devil is on the rampage, Derwin says. In the Garden of Eden, "there wasn't anything hurting your environment. We'll probably never see the bayou like God made it in the beginning until He fixes it himself. And that will happen pretty shortly, so it don't matter how much man destroys."[69]

"*It don't matter how much man destroys*" is a naive way to dismiss causes of climate change, echoing statements by politicians who say, "only God can destroy the earth," or it is blasphemous to contend that man can destroy the earth.[70] It is not my intent to question religious beliefs, but it seems that those who say these things have a willful ignorance or an expedient dismissal of troubling aspects of climate change. In some religions apocalypticism uses millenarist beliefs to justify dismissal of climate science. These beliefs are based on prophecy of "end times," the transformation of the world through divine intervention that will change everything in the near future. Since the millenarist has few concerns about preserving the present world, any concern about climate change can be discarded.

William Vollman's interviews with Appalachian coal miners' families elicited a similar reaction, when he asks about climate change: "I don' know if I believe in that, Bill… [the earth] works the way God intended. It's designed for perfection, and He's gonna do what He's gonna do."[71] This expression of fatalism displays the form of ideology that permits the speaker to avoid confronting the implications of climate science. While we should not disparage this way of thinking among people who may depend on coal for their livelihood, it shows the difficulty of overcoming beliefs that lead to collapse ideology.

Beliefs such as this are based on religious interpretations of nature, such as God created earth and made it congenial for human society. They depend on human experience with a specific epoch, the Holocene climate, that has been stable for over 10,000 years. As Ian Angus observed, "The idea that the natural world is fundamentally stable and unchanging has a long history. In its oldest version, it is religious: God created a perfect world, and if humans disturbed that perfection, God would in time restore it."[72] The natural world is not as stable and unchanging as believed; there are many periods in the distant past when temperatures and conditions fluctuated much more than the recent past. Nevertheless, some ideologues will insist that only God can make dramatic changes so any findings of climate science that suggest otherwise are heretical.

Arguments about the supremacy of the divinity are used by ideologues such as Bastardi to criticize climate scientists and activists of "playing God."

> Trying to control nature and pretending you are the master of Creation set man up as god. Because if man can control what God created, is not man greater than God? If that is the case, then men who see themselves as god seek to control those under them.[73]

With Bastardi's assertion, the argument shifts from humans defying God to humans playing God in order to control people. This argument is a kind of counter-elitism, accusing climate scientists and activists of *"controlling those under them"* as if climate science is some kind of theology.[74]

Sometimes collapse ideologues will accuse climate activists of witchcraft: Morano says, "The notion that a UN agreement to limit emissions will somehow alter the Earth's temperature or storminess borders on belief in witchcraft."[75] This is redolent of the argument that nothing we will do will affect climate, for good or bad. Therefore, if we believe that we can control climate by limiting emissions, we are engaging in witchcraft. While this may sound like a rather frivolous form of argument, at least one author has accused climate activists of Satan worship.[76] I would like to think that we can ignore this kind of activist-bashing, but it does tend to reinforce public views that dismiss climate science, and justifies living standards based on thoughtless consumption of fossil fuels.

Perhaps the most egregious example of climate injustice is the intergenerational injustice of people living today usurping the resources that future generations will need to live. Naomi Klein has put the issue succinctly: "Because one of the most unjust aspects of climate disruption (and there are many) is that our actions as adults today will have their most severe impact on the lives of generations yet to come, as well as kids alive today who are too young to impact policy—kids like Toma [her son] and his friends, and their generation the world over."[77]

One of the unfortunate paradoxes of climate is that those responsible for problems leave it to their children and grandchildren to deal with the outcomes. As Klein notes, "These children have done nothing to create the crisis, but they are the ones who will deal with the most extreme weather – the storms and droughts and fires and rising seas – and all the social and economic stresses that will flow as a result. They are the ones growing up amidst a mass extinction, robbed of so much beauty and so much of the companionship that comes from being surrounded by other life forms."[78]

Intergenerational injustice is insidious because it relies on moral sensitivity to the effects of current generations on future generations. Since those future generations are not yet present, or are too young to protest, they are going to pay for climate costs that they have not caused. Adults today are, perhaps unconsciously, condemning their descendants to a bleak future. In France, this became explicit when demonstrators fought a carbon tax. "'Why should we have to finance their projects?' [a demonstrator] said, referring to the government's plan to discourage car use through gas taxes. Many in the crowd said that they did not disdain the government's environmental goals, but that their own survival was more important."[79] The argument that *"their own survival was more important"* is a prime example of intergenerational injustice – disdaining the "environmental goals" of the government that would reduce the damage to future generations by requiring some sacrifices of the current generation. Survival of humanity is put in peril by this generation's concept of its survival.

French demonstrators were also using a short-term view of economic life in fighting a carbon tax. "Mr. Macron is concerned about the end of the world, while they are worried about the end of the month."[80] It is hard to argue with people who are desperate about their economic situation on a month-by-month basis, but their eventual fate is more damage from climate change than from a carbon tax.

One should not reason from the French experience that poor people are not concerned with climate change. Sociologist Neil Gross explained that "in many of today's capitalist democracies, class and status resentments, fostered by rampant inequality and whipped up by opportunistic politicians, have developed to such an extent that issues like the environment that affect everyone are increasingly seen through the lens of group conflict and partisan struggle."[81] Motivated by partisan views of the situation, voters in many countries have defeated efforts to address climate change, not because they are indifferent to the issue but because of their affiliations.

Climate change is a "wicked" problem that exacerbates the French situation. It is understandable that French citizens would oppose a carbon tax. Their society, like so many around the world, depends on fossil fuels because of a legacy of rural development and urban sprawl promoted by auto use. Most residents have become dependent on private cars for transportation. Many of the rioters in France came from rural areas into Paris to protest the threat to their livelihoods from the carbon tax. The government was forced to cancel the tax, even though many economists and scientists believe it is the best way to address climate change.

Transportation issues that impact on climate change stem from the issue of mobility. As French sociologist Alexis Spire noted, "As small businesses have been dying in these smaller cities and towns, people find themselves forced to seek jobs elsewhere and to shop even for basic goods in malls. They need cars to survive, because regional trains and buses have declined or no longer serve them. Once you begin to unpack the Yellow Vest phenomenon, the uprising is a lot about mobility."[82] While it is well established that everyone pays for climate effects of transportation, it is not a simple equation when the need for mobility is taken into account.

It has become increasingly obvious that the real cost of fossil fuels is much higher than the market prices. One clear sign of climate science dismissal is the refusal to include those costs in economic systems. The question I have is why the persistence of collapse ideology – are believers unaware of the consequences of delaying action? The issue of transportation is a complex issue, and we cannot easily dismiss people as shortsighted because of their mobility needs. Nevertheless, people have become dependent on destructive forms of energy.

Energy Fetishism Redux

As noted earlier,[83] the progression of beliefs about lifestyles and the role of modern humans in society have led to a fetishism of energy. This fetishism starts from a reliance on fossil fuels and leads to fear about deprivation if fossil fuels are not available. A consequence of the fetishism of energy is collapse ideology itself, which feeds the inability of people to see beyond their immediate needs for fossil energy to a future in which those needs can be met with renewable energy. This blind spot in the human psyche is impeding the change that is needed to avoid climate catastrophe. Opening people's eyes to new methods of fueling their lifestyles is hampered by a fixation on present forms of energy to the exclusion of alternatives.

A prime example of energy fetishism is the automobile. While it has enhanced mobility with freedom of movement, it carries a major cost: up to 40% of carbon emissions in some areas. The U.S. government under the Obama Administration attempted to deal with this problem by raising mileage standards and encouraging alternatives to the internal combustion engine. The efforts to address the problem by the Obama Administration have been attacked by the Trump Administration, which has proposed rolling back the mileage standards (from 54 miles per gallon to 37 mpg). Interestingly, even the auto industry dislikes the attacks. They reduce the certainty of standards, since many states will try to maintain the higher standards and auto companies do not like dealing with different standards for different markets. Reportedly the Trump Administration will attack state standards (led by California and followed by 13 other states) in court, but court cases can take years.[84] It is likely that the auto companies will have to meet the higher standards despite the policies of the Trump Administration, driven by collapse ideology.

Of course, the fact that California and other states might use mileage standards as a policy instrument against climate disruption arouses the ire of collapse ideologues such as Heartland Institute, which said "For decades, multiple administrations in Washington, DC allowed California to dictate environmental mandates to the rest of us. Revoking Obama's draconian fuel mileage restrictions, and California's dubious exemption to federal standards, would make cars safer and more affordable."[85] It seems inevitable that these ideologues would attack any state that presumes to address climate change. The argument that larger, heavier cars are safer and more affordable is tenuous, assuming that automobile companies can only make one-dimensional changes in their technologies. Many changes in the design of cars have made them more efficient *and* safer, through the use of crumple zones, aluminum, computerized fuel injection and hybrids, for example.

One of the tendentious arguments used by the Trump Administration, reflecting claims by collapse ideologues, is that the higher mileage standards would avoid only 0.64 parts per million (ppm) of carbon concentrations (which are now at 410 ppm) by 2100. This argument demonstrates two problems of collapse ideology: understating the real effects of fossil fuels on the climate[86] and treating the U.S. in isolation from other countries. Even if U.S. emissions are lowered by a small amount, we are part of a worldwide system of energy use and all nations must participate in the effort to lower emissions. As we will see below, the U.S. has isolated itself by denouncing the Paris Agreement, but that does not mean we are off the hook. We will definitely suffer as much as other countries even the U.S. tries to avoid limits to carbon emissions.

With regard to the actual emissions affected by these rule changes, data from two sources indicate that they would increase substantially. "Weaker vehicle rules could add an extra 28 million to 83 million metric tons of carbon dioxide to the atmosphere in 2030 alone, according to a Rhodium Group analysis. The proposal to relax rules for coal plants would add another 47 million to 61 million tons of CO_2 that same year, according to the E.P.A.'s own numbers."[87] That is more carbon emissions than either Belgium or Greece emit from all sources in one year.

Wildfires, Storms and Floods

President Trump has used his own version of reality in framing causes for climate change consequences such as wildfires. Trump tweeted "California wildfires are being magnified & made so much worse by the bad environmental laws which aren't allowing massive amounts of readily available water to be properly utilized. It is being diverted into the Pacific Ocean. Must also tree clear stop fire spreading!" He was apparently referring to California's policies on water flow that keep some streams flowing, when drought reduces them, so that fish can survive. His tweet was countered by a statement from Scott McClean, deputy chief of the California Department of Forest and Fire Protection: "We're having no problems as far as access to water supply. The problem is changing climate leading to more severe and destructive fires."[88] Trump's suggestion that loggers clearcut forests to avoid fires is a naïve misunderstanding of forestry. He has also suggested, referring to the Paradise fire, that forest managers should "rake" the forest floor, just as Finland has done. (Finnish officials denied that claim.)

Trump's reaction to the 2018 report of the IPCC (Intergovernmental Panel on Climate Change, see Appendix) demonstrates his naivete on most climate issues. First, he questioned the legitimacy of the IPCC: "I want to look at who drew it, you know, which group drew it." Then he suggested that there are other groups that would prove the opposite and that other reports were as valid as the UN's, though he did not specify any. "I can give you reports that are fabulous, and I can give you reports that aren't so good," Trump said.[89] Where he gets the idea that there are reports on the climate as "fabulous" is unknown, but it might involve the Heartland Institute, which produces periodic reports attacking the IPCC findings.[90] He echoed some of the Heartland skepticism: "I can also give you reports where people very much dispute that, you know, you do have scientists that very much dispute it." Trump seems to think he can make up his own "science" about climate change: "Brutal and Extended Cold Blast could shatter ALL RECORDS - Whatever happened to Global Warming?"[91]

The idea that Trump would dismiss the findings of thousands of scientists, compiled in reports vetted by many experts, is an indication of collapse ideology. Obviously, he does not want to face the reality of climate change that has already displaced many people and damaged property. The

New York Times has reported data that shows, "On average, floods and storms have displaced nearly 21 million people every year over the last decade, according to the Internal Displacement Monitoring Center. That is three times the number displaced by conflict.

Worldwide, according to Munich Re, damaging floods and storms have more than tripled in number since the early 1980's. There economic losses have risen sharply with two record years in the last decade in which damages topped $340 billion."[92] In one year, 2014, 17.5 mission people were displaced by disasters related to climate, more than ten times the 1.7 million displaced by "natural" disasters such as earthquakes or volcanoes.[93] These facts demonstrate that the threats projected by the IPCC (see Appendix) are already occurring, and creating refugees, but Trump and other politicians would like to ignore them.

How can humans invent ideology that permits them to dismiss reality and continue on a self-destructive path? Yuval Harari has called humans the *post-truth species*. "Homo sapiens is a post-truth species, whose power depends on creating and believing fictions. Ever since the Stone Age, self-reinforcing myths have served to unite human collectives. Indeed, Homo sapiens conquered this planet thanks above all to the unique human ability to create and spread fictions. We are the only mammals that can cooperate with numerous strangers because *only we can invent fictional stories, spread them around, and convince millions of others to believe in them.* As long as everybody believes in the same fictions, we all obey the same laws and can thereby cooperate effectively."[94] While cooperating effectively on some activities may be functional for humanity, cooperating effectively on increasing use of fossil fuels is highly dysfunctional.

As the ultimate form of collapse ideology, studied indifference to and willful ignorance of natural limits to human activities is the force that will drive the climate to extremes. Continued use of fossil fuels as if there is no tomorrow is a sure sign of ideology leading to collapse. As Joel Wainwright has observed, "It is not hard to find evidence of this reactionary tendency today, epitomized in the continued influence of climate science denial in mainstream political discourse, especially in the United States. The millenarian variety of this formation embraces an ideological structure that renders it impervious to reason. Indeed, that is the point."[95]

Willful Ignorance

"*Impervious to reason*" is indeed the point. We adhere to collapse ideology, not because we think through the implications of our beliefs for our behavior and our future, but because we need to cling to something that justifies our lifestyle. The rejection of climate science is a not always a deliberate, conscious effort to avoid the implications. One author, Robert Lifton, has even called it "climate change rejection."[96] Lifton goes on to note that to avoid truths that would violate their own and their sponsors' individual and group identity and ideology, people look to other sources for climate change such as religion. "Man cannot destroy the world, only God can"[97] is the mantra for this way of thinking. As Lifton concludes, this enables people to "divest themselves of a sense of responsibility." It must be very comforting for people to think that they are not responsible for climate change even as they consume increasing amounts of fossil fuel.

Sometimes popular reactions to climate change go beyond the passive-aggressive form described above. When France imposed fuel taxes as a mode of climate action, the streets erupted with violence, as noted above. Much of the popular reaction was inflamed by the use of social media, which led the *New York Times* editorial board to condemn the violence: "But the power of social media to quickly mobilize mass anger without any mechanism for dialogue or restraint is a danger to which a liberal democracy cannot succumb. Mr. Macron and the Parliament were democratically elected only 18 months ago, and the reforms they have been pursuing, both within France and in the European Union, and on the environmental front, were what they openly promised in those elections and what France needs."[98]

Migration and Climate Injustice

Droughts have a way of driving people off the land and into cities, where they experience violence and other problems that impel them to migrate. The "caravan," a group of thousands of

Central American migrants, made its way from Honduras and other countries to the U.S. border in 2018, where they were met by stiff resistance. Ideally, they would be considered refugees: "If your farm has been dried to a crisp or your home has been inundated with water and you're fleeing for your life, you're not much different from any other refugee. The problem is that other refugees fleeing war qualify for that status, while you don't," said Michael Doyle, an international relations scholar at Columbia University.[99] The effect of this kind of migration is tension and even civil war, as in the case of Syria. There is a great deal of internal displacement from climate change, and increasingly these refugees are moving across national boundaries. Climate refugees are not recognized in international law as other refugees are, so they often find crossing borders difficult.

Climate change is not the only cause of forced migration, but it is increasingly involved in the movement of people due to changes in their environment. ""The world today is witnessing an era of unprecedented human mobility with more than one billion people on the move," said Sarat Dash, the International Organization for Migration chief of mission in Bangladesh. "Forced migration due to poverty, conflict, climate change, and disasters can lead to deterioration in development outcomes." For South Asia, he said, "the range of sudden and slow onset events like changing rainfall, rising sea-levels, coastal erosion, floods, salinity intrusion and droughts put communities at greater risk impacting their economic, health, food and security conditions."[100] These conditions are exacerbated by climate change, and it is increasingly likely that they will become worse as the world warms.

One of the more difficult aspects of climate action is the division of the world into different political jurisdictions. Even in the U.S., there are different approaches to climate action in states, with more or less effective policies. California, for example, has strong climate policies in place while next door in Nevada and Arizona policies are not as strong. The U.S. Climate Alliance is a group of 17 states and a number of cities and counties, but it covers about 40% of the economic activity of the U.S. and only 35% of the emissions.[101] The Alliance, however, has made pledges for reducing annual carbon emissions by 1.5 to 2.2 gigatons of carbon dioxide equivalent, twice as much as Canada's annual emissions.[102] This would reduce emissions by the amounts pledged in 2015 by the U.S. during the Obama Administration, despite the backsliding of the Trump Administration.

Action on environmental problems has long been a problem of jurisdictional control. It is also a problem of environmental injustice, as some states and localities are healthier than others, depending on how well the political system addresses environmental problems.[103] In the case of climate change, this problem is writ large, because carbon dioxide and other greenhouse gases do not respect national boundaries. The problem is exacerbated by the fact that some countries are economically and technologically better equipped to address climate change than others.

One indigenous leader in Colombia, Fany Kuiru Castro, noted, "The best land has been taken by the ranchers and the palm oil and sugarcane companies. There are thousands of cases where companies are in conflict with indigenous communities over land. This is an urgent issue that needs to be resolved."[104] Indigenous communities are most effective in preserving tropical rainforests, and to recognize the jurisdictional rights of indigenous peoples will maintain the function of tropical rainforests in carbon sequestration. Brazil's new president, Jair Bolsonaro, has proposed opening more of the Amazon rain forest to development, which encroaches on indigenous communities and threatens its function as the "lungs of the planet."

One of the problems that jurisdictional control by nation-states poses, to the world's climate, is the "free-rider" problem. Some countries, notably the U.S., can avoid costs of the obligations of the Paris Agreement by denouncing it, but the world's climate will be worse for everyone even if all other parties to the agreement meet their obligations. Greenhouse gases do not recognize national sovereignty.

Ironically, Donald Trump invoked this "free-rider" argument when he argued against U.S. support of the Paris Agreement:

> The Paris Agreement is fatally flawed because it raises the price of energy

for responsible countries while whitewashing some of the worst polluters in the world. I want clean air and clean water and have been making great strides in improving America's environment. But American taxpayers – and American workers – should not pay to clean up other countries' pollution.[105]

Where does he get the idea that the Paris Agreement lets anyone off the hook? While compliance is voluntary, it does require all nations to submit "Nationally Determined Contributions" and make their accounting of carbon emissions public. Developing countries are mentioned as possible recipients of aid in accounting and reducing their emissions, but they are not exempted. Of course, Trump has reneged on the $2 billion pledged by the U.S. to aid in these efforts, so he has little basis for claiming that they are "whitewashing" emissions.

Acknowledgement of the developed-developing country asymmetry of climate calls for actions at the global level, because much of the climate justice issue involves addressing harm to less developed countries who are responsible for only small amounts of carbon emissions. This justice issue was addressed in the Paris Agreement. Text of the agreement included references to "common but differentiated responsibilities" and "support of developing country Parties" as examples of climate justice:[106]

- "This Agreement will be implemented to reflect equity and the principle of common but differentiated responsibilities and respective capabilities, in the light of different national circumstances." (from Article 2)
- "The efforts of all Parties will represent a progression over time, while recognizing the need to support developing country Parties for the effective implementation of this Agreement." (from Article 3)
- "Parties aim to reach global peaking of greenhouse gas emissions as soon as possible, recognizing that peaking will take longer for developing country Parties." (from Article 4)[107]

These provisions of the Paris Agreement and previous agreements such as the Kyoto Protocol codify the impetus for climate justice among diplomats. Unfortunately, provisions for climate justice do not match political realities in the U.S. They have caused political problems for U.S. support of the UN Framework Convention on Climate Change (UNFCCC). Along with denouncing the Paris Agreement, the Trump Administration has also threatened to withdraw the U.S. from the UNFCCC itself, which was signed by President George H.W. Bush and ratified unanimously by the U.S. Senate. All climate agreements recognize the different responsibilities of industrial countries, which have contributed the most to cumulative carbon emissions, and other countries that have contributed little. While this makes sense as a form of climate justice, it is not popular with U.S. politicians, who have used it as an excuse for rejection of agreements such as the Kyoto Protocol and the Paris Agreement.

Climate justice issues also arise when climate crises such as wildfires arouse questions of fairness and inequality. Some of the celebrities in Malibu, California, hired private firefighters to protect their homes from a wildfire. In one response, the issue was framed as a crisis of capitalism: Muquing Zhang said, "We are entering a dystopian future in which class-privileged white people are using privatized systems and their obscene wealth to avoid the catastrophic environmental effects of the racist capitalist system that they forced upon the world.... These fires and their disparate impact demonstrate a necessity for structural action on climate change and support for those who will be most harmed – poor people and people of color."[108] Although somewhat radical in its analysis of climate justice, it identifies some of the areas where future conflicts are likely to occur. Will wealthy people be able to escape the effects of climate change? Will they be able to avoid its consequences while others suffer? Although private firefighters may stave off the immediate crisis, in the long term all of us will be threatened by climate disruption.

Sabotaging the Planet

> *The possibility of rapid, global carbon mitigation as a climate change abatement strategy has passed. The world's elites, at least, appear to have abandoned it – if they ever took it seriously.*
> *Wainright and Mann[109]*

Many countries in the world – 197, virtually all the UN members – have seen the need for climate action and agreed on the Paris Agreement. Only one – the U.S. – has reneged on its commitment. The Paris Agreement was written in a way that requires four years' wait for the U.S. to actually withdraw, and that deadline falls on the day after the 2020 presidential election. It will be interesting to see if the Trump Administration changes its stance after several years of record heat, floods and wildfires. When Trump leaves office, the next president may find it compelling to reengage the U.S. in the Paris Agreement.

Although my book *Sabotaging the Planet* was published before the 2016 U.S. election, the inclinations toward sabotage of the Paris Agreement were already apparent in the U.S. Congress. In the presidential election, the candidate who called climate change a hoax won, and withdrew the U.S. from the agreement.

President Donald Trump cited inaccurate data when he announced the withdrawal of the U.S. from the Paris Agreement on June 1, 2017.

Jobs and Economic Growth

Trump stated that the Paris Agreement would mean "lost jobs, lower wages, shuttered factories, and vastly diminished economic production." He used negative statistics about the economic impact from the climate deal, including a $3 trillion drop in gross domestic product, 6.5 million industrial sector jobs lost, and 86 percent reduction in coal production, all by 2040.

Trump used data from an unreliable source to make his case.[110] His data makes worst-case assumptions that may inflate the cost of meeting U.S. targets under the Paris Accord, while largely ignoring the economic benefits to U.S. businesses from building and operating renewable energy projects.[111]

The myth of economic damage from climate action pervades in collapse ideology. It is a rationalization based on fear of change, that a low-carbon economy will somehow be a poorer economy with fewer jobs and a meager lifestyle. In fact, jobs increase with expansion of renewable energy. There are already twice as many jobs (374,000) in renewable energy fields than in coal and oil combined (187,000).[112]

Growth in renewable energy will mean more jobs, but if that growth is not managed well there will be economic decline.[113] People tend to fear the possibilities of losing employment. Job transition is a major element of a just policy on climate change.

Fear of change is pervasive in climate science dismissal. It is not only a deterrent to understanding and acting on climate science, it can also paralyze people who do understand climate science but do not want to act on it. As religious leader Mishra-Mazetti has observed, "In looking at the different ways people respond to the concepts of climate change, ecological damage, and environmental injustice, we can see that much of it is a reaction to fear: fear of having to change how we think about the world and our lives, fear of catastrophe, fear of responsibility for actions and choices." [114] Unfortunately, fear is not a positive influence and tends to paralyze action, not motivate it.[115] President Trump counts on fear when framing his policies, as has been well documented by Bob Woodward. It is this fear that Trump relies on when he claims that addressing climate change will cost jobs and reduce economic growth.[116]

Reducing Carbon Doesn't Affect Temperatures

Trump said, "Even if the Paris Agreement were implemented in full, with total compliance from all nations, it is estimated it would only produce a two-tenths of one degree... Think of that, this much...reduction in global temperature by the year 2100. Tiny, tiny amount."[117]

Research cited by Trump was outdated, and the researchers said that the valid number would be one degree Celsius, not two-tenths.[118] Even this is not enough, but it is five times what Trump claimed.

Even the temperature limits in the Paris Agreement are not likely to be effective. The 2C (two degrees Centigrade) limit is already at a dangerous level; as climatologists Keven Anderson and Alice Bows-Larkin observe, 2C "represents a threshold, not between acceptable and dangerous climate change, but between dangerous and 'extremely dangerous' climate change."[119] Societies are going to suffer high levels of starvation, inundation, drought and other effects as a result of the 2C limit; that is why the Paris Agreement also includes aspirations to limit temperature to 1.5C.

Even Paris limits are still not enough. "There is increasing and very robust evidence of truly severe and catastrophic risks even at the lower bounds of these temperature targets," said Peter Frumhoff, director of science and policy at the Union of Concerned Scientists.[120] He described some of the effects at the 1.5C level as disastrous, and going above the 2C level would be catastrophic. "But while we might call 2C an upper bound, let's not pretend that we're on a 2C path – we are way above that," he said. The UN Intergovernmental Panel on Climate Change (IPCC) reported in 2018 that the world must start reducing emissions now and reach zero net emissions (emissions minus sequestration) by 2050, as envisioned in the Paris Agreement. (*See Appendix*) While scientists are warning of climate consequences even if they stay within the Paris limits, societies are not responding well.

While limits identified in the Paris Agreement or IPCC reports are necessary, they are currently infeasible given the low level of climate action that is currently underway around the world. The irony is that infeasibility in reaching Paris goals is used by some as an excuse to do nothing, rather than gearing up to do what is absolutely necessary.

An ineffectiveness myth embodies much of the rationale found in collapse ideology. For example, the Heartland Institute asserts that climate change is natural, not human-induced, and that any human actions (good or bad) will have miniscule effects on temperature. Another version of this myth is that severe climate change already existed before human development, a truism that conflates natural and human causes. The myth is also is embodied in the assertion that there has been a hiatus in global warming during the past two decades, since 1998. In other words, Trump and believers in collapse ideology assert that carbon emissions have no effect on global temperatures and that reducing them will have no effect on temperature rises. They cite the fact that carbon emissions have been steadily rising during the past two decades, while temperatures remain stable. This is called the "pause" myth.

The "pause" myth is based on cherry-picking data. Beginning temperature data with 1998 uses an especially high temperature record year as a baseline (1998 was an El Nino year, with the highest recorded global average), then picking subsequent years with lower global averages for comparison. It is also based on the use of defective satellite measurements. Data cited by John Christy and associates at the Earth System Science Center at the University of Alabama in Huntsville are corrupted by the deterioration of orbits of the satellites.[121]

The best refutation of the "pause" myth is the NOAA measurements of actual temperature increases. NOAA data clearly indicate an upward trend since 1994, with only 1998 showing much deviation from that trend. Nevertheless, perpetrators of the climate myth that temperature increases have paused use 1998 as a baseline, to claim there is no warming, despite increasing concentrations of carbon dioxide. This is an unethical manipulation of data and an immoral manipulation of facts. It purposely ignores the highest recorded global temperature data in 2005, 2009, 2010, 2013, 2014, 2015, 2016 and 2017, all above the 1998 record. In the following graph, the years are arranged in order of increasing anomaly from the long-term average, with 1998 ranking ninth out of ten.[122]

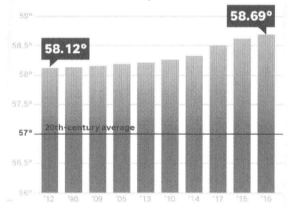

Another factor overlooked by perpetrators of the myth is that 90 percent of warming goes into the ocean.[123] Because water is able to absorb large amounts of heat, it retains a lot of the temperature increases from global warming. There is a downside to this aspect of climate change, however: As the ocean warms, it expands, raising sea levels and impacting coastal communities. Warm water is also less likely to absorb and hold carbon dioxide, so it returns some CO_2 to the atmosphere, accelerating the buildup of carbon concentrations.

Politicians will sometimes try to explain sea level rise with causes unrelated to climate change. At a hearing of the House Committee on Science, Representative Mo Brooks of Alabama got into an argument with Phil Duffy, the president of the Woods Hole Research Center and former senior adviser to the US Global Change Research Program.

"Every time you have that soil or rock deposited into the seas, that causes the sea level to rise, because now you've got less space in those oceans, because the bottom is moving up," Brooks explained to the scientist. Duffy tried to interject but was unsuccessful. "What about the white cliffs of Dover, California, where time and time again you're having the waves crash against the shorelines, and time and time again, you're having the cliffs crash into the sea. All that displaces water, which forces it to rise, does it not?" Brooks continued. "I'm pretty sure that on human time scale, those are minuscule effects," Duffy replied calmly.[124]

How many rocks would have to fall into the sea to cause higher water levels around the globe? Brooks does not know what he is talking about. With this low level of sophistication, there is little wonder that politicians often get the science wrong.

Saving Coal Jobs

President Donald Trump said one reason he withdrew his country from the Paris Agreement was to save coal jobs, but the move is unlikely to revive the U.S. coal industry. Coal-fired power generation in the United States is expected to fall by 51 percent by 2040, with a 169 percent increase in renewable power helping to fill the void.[125] Nearly all coal station shutdowns are due to economic factors such as the lower price of natural gas and declining cost of renewables, not regulations.

EPA rules on coal proposed by the Obama Administration have been rewritten by the Trump Administration to reduce the burden on coal plants. This means that they can now "experience an increase in annual emissions due to increases in operation" according to EPA's own legal notice.[126] The new rules have not gone into effect because of court challenges. Whether they will ever have a climate effect is an open question, due to the likelihood that market forces – principally the lower cost of natural gas – are likely to push coal out sooner than any regulations.

Relying on the myth of saving coal jobs is a tactic of sectorial politics; it appeals to a segment of the population impacted by social change by blaming some other sector or elite group. Trump did not blame the natural gas industry for loss of coal jobs; instead he implied that changes compelled by the Paris Agreement are responsible. Even if coal production revives somewhat, employment will not increase much because of market forces that have reduced coal jobs. Substitution of gas for electrical generation and automation of mine jobs have already reduced coal jobs by 70% from 1986 to 2018.

Even Trump's EPA, headed by a former coal lobbyist, admitted that the Trump plan for coal-fired power plants is not good. It would cause 1400 premature deaths, in contrast to the Obama Clean Power Plan that would have saved 3600 lives from premature death. The Trump plan will also result in 48,000 cases of aggravated asthma, while the Obama Clean Power Plan program would have prevented 180,000 fewer missed school days per year by children with ozone-related illnesses. This is all on top of the approximately 600 megatons of carbon per year that will be emitted if of Trump's wishes are implemented.[127]

Trump's Fantasies

It should not come as a surprise that Trump is against the Paris Agreement – he is confrontational and unwilling to work with other countries on an equal basis. His zero-sum view of international politics precludes full cooperation with the international community and his populist approach to politics means he will pander to a base that is not favorable to international cooperation. A business magazine editorial noted: "It should be obvious that this form of politics is antithetical to global decarbonization and climate action. Its inherent divisiveness is the precise opposite of the sense of shared endeavor and international co-operation contained in the Paris Agreement. No wonder Trump hates the treaty so much — it is the diametric opposite of everything he believes."[128] Only collapse ideology could lead one to denounce international cooperation on the most significant threat faced by humanity.

President Trump relies on a principle of propaganda best enunciated by Harari: "In fact, false stories have an intrinsic advantage over the truth when it comes to uniting people. If you want to gauge group loyalty, requiring people to believe an absurdity is a far better test than asking them to believe the truth."[129] The absurdity that Trump is asking us to believe is that climate is not something to worry about, and that we can ignore the efforts of the international community to counter climate disruption.

What can be done about a president who uses climate myths to justify his denunciation of the Paris Agreement? Many suggestions about reining him in or impeachment have been made but few are realistic. A more plausible scenario is defeat in the 2020 election after climate change becomes so palpable that voters cannot ignore it. Columnist Thomas Friedman has outlined this scenario:

> …if in 2020 we're in the midst of even more damaging droughts and storms than we are today, Democrats may be able to run against Trump's make-America-polluted-again environmental strategy and his refusal to either acknowledge the threat of climate change or seize the incredible opportunity it offers America to become richer, healthier, more secure and more respected by leading the world in clean energy technologies. Trump has no answer for that. He doesn't believe the

climate science that NASA is telling him is true. He is trying to bring back coal precisely when wind, solar and efficiency are becoming cheaper, cleaner, healthier alternatives – precisely when four of the five biggest wind states are red states and precisely when China has committed itself to owning the clean power and electric car markets of the future![130]

Politicians are notoriously reticent to change what they believe to be a winning strategy but Trump may be forced to either change his climate policies or leave office in disgrace. I would wager on the latter given Trump's stubbornness and the support of his administration for collapse ideology.

Trump continues to sabotage the planet, in the years after denouncing the Paris Agreement. In June 2018 he attended a G7 meeting in Canada, during which a statement on climate change was adopted by six of the seven, with a dissent by the U.S.:

> Canada, France, Germany, Italy, Japan, the UK and the EU agreed new language on the importance of carbon pricing and a "just transition" to clean energy, as well as reaffirming their commitment to the Paris Agreement.[131] The US asserted its position in a separate paragraph, prioritising economic

growth and energy security. It would support countries in using "all available energy sources", including to "access and use fossil fuels more cleanly and efficiently."

Even the G-7 statement was too much for Trump; he tweeted his rejection of the entire statement with a complaint about trade: "I have instructed our U.S. reps not to endorse the communiqué." The U.S. position favoring "access and use fossil fuels more cleanly and efficiently" is the standard Trump jargon for promoting coal and oil. The Trump administration has been promoting coal-fired power plants even when utilities want to shut them down, citing national security reasons for requiring power plants to maintain a 60-day supply of fuel. Even on the face of it, this is a ridiculous rule; how can gas-fired power plants maintain 60 days of natural gas? One could argue that solar and wind generating facilities have an infinite supply of fuel, so they surely would qualify.

Trump's Cabinet

Mike Pompeo was appointed secretary of state by Trump in 2018. He had been a member of Congress when the Paris Agreement was signed, and he called it a "costly burden" that imposed unfair costs on the U.S. He described former President Obama's support of the agreement as "bow(ing) down to radical environmentalists," and he blasted Obama for a "perverse fixation on achieving his economically harmful environmental agenda" in the Paris negotiations.[132] His dismissal of climate science has been somewhat nuanced; when confirmed in 2017 as CIA Director, he said that he would not comment on climate change because it is not a national security issue. This contradicts Defense Department findings that it is a national security threat.

Brock Long, Director of the Federal Emergency Management Agency (FEMA), has equivocated on climate change, which is one of the contributing factors to the emergencies with which he must deal. He said "The term climate change has become such a political hot button that, I think, I keeps us from having a real dialogue,"[133] a clever spin on the issue. He is saying, in effect, that politicians do not want to deal with climate science because it could require them to defy special interests or to stop pandering to their base. In this respect, Long joins a long list of politicians who equivocate on climate by disavowing responsibility for grappling with climate science.[134]

The avoidance of climate action is built into the political system. As suggested by Nathaniel Rich, "Political problems had solutions. And the climate issue had none.

Without a solution – an obvious, attainable one – any policy could only fail. No elected politician desired to come within shouting distance of failure. So when it came to the dangers of despoiling our planet beyond the range of habitability, most politicians didn't see a problem."[135] If they don't see the problem, or don't want to see a problem, it is easier to deny the need for action – a sure sign of collapse ideology.

According to Climate Solutions' analysis, more than half of the Trump cabinet denies climate science or considers it unnecessary to use in policy decisions. According to an analysis by the Center for American Progress Action Fund, if a full meeting of President Donald Trump's cabinet took up the issue, most of the members would dismiss climate science.[136]

And what about Vice President Mike Pence? He is right there with the worst of them. Before becoming vice president, he had opportunities as a member of Congress to reduce dependence on fossil fuels, after Hurricane Katrina made it obvious something had to be done. "Though climate scientists have directly linked the increased intensity of hurricanes to warming ocean temperatures, that didn't stop Pence … from calling on Congress to repeal environmental regulations on the Gulf Coast, give permission for new oil refineries in the United States, and green-light drilling in the

Arctic National Wildlife Refuge."[137] All of these policies contribute to climate change, but collapse ideology justifies them.

By abandoning climate action, Trump, Pence, Pompeo and others in the Trump Administration are perpetuating climate myths that relieve them of difficult decisions about reordering the economy to fossil-fuel free energy. All of these myths and many others persist in the political sphere because it is in the interests of some politicians to pander to ignorance and feed preconceptions about maintaining current energy systems and lifestyles based on them.

Congress

Executive branch politicians are not the only ones to perpetuate myths of climate denial. The U.S. House of Representatives passed Resolution 119 on July 19, 2018:

> Expressing the sense of Congress that a carbon tax would be detrimental to the United States economy.
>
> Whereas a carbon tax is a Federal tax on carbon released from fossil fuels;
>
> Whereas a carbon tax will increase energy prices, including the price of gasoline, electricity, natural gas, and home heating oil;

Whereas a carbon tax will mean that families and consumers will pay more for essentials like food, gasoline, and electricity;

Whereas a carbon tax will fall hardest on the poor, the elderly, and those on fixed incomes;

Whereas a carbon tax will lead to more jobs and businesses moving overseas;

Whereas a carbon tax will lead to less economic growth;

Whereas American families will be harmed the most from a carbon tax;

Whereas, according to the Energy Information Administration, in 2016, fossil fuels share of energy consumption was 81 percent;

Whereas a carbon tax will increase the cost of every good manufactured in the United States;

Whereas a carbon tax will impose disproportionate burdens on certain industries, jobs, States, and geographic regions and would further restrict the global competitiveness of the United States;

Whereas American ingenuity has led to innovations in energy

exploration and development and has increased production of domestic energy resources on private and State- owned land which has created significant job growth and private capital investment;

Whereas United States energy policy should encourage continued private sector innovation and development and not increase the existing tax burden on manufacturers;

Whereas the production of American energy resources increases the United States ability to maintain a competitive advantage in today's global economy;

Whereas a carbon tax would reduce America's global competitiveness and would encourage development abroad in countries that do not impose this exorbitant tax burden; and

Whereas the Congress and the President should focus on pro-growth solutions that encourage increased development of domestic resources: Now, therefore, be it

Resolved by the House of Representatives (the Senate concurring), That it is the sense of

Congress that a carbon tax would be detrimental to American families and businesses, and is not in the best interest of the United States.

Passed the House of Representatives[138]

Congress perpetuates the myth that "pro-growth" solutions require the use of fossil fuels, and that a carbon tax would hurt U.S. families and businesses. It assumes that growth cannot be separated from fossil fuel use, and that low-income communities would be hurt by a carbon tax. Both of these assumptions are promoted by those who would deny the advantages of renewable energies for the economic growth, including the economic growth of low-income communities.

Ideologues Rule

Myths perpetuated by politicians are echoed in the climate dismissal press. While some ideologues are edging toward admitting a human role in climate change, they continue to deny actions that may resolve the problem. Note the wording in a Fox News article: "The Paris accord is a bad deal because even if greenhouse gases really are a huge threat, this treaty wouldn't do much about them.[139]" *"Even if greenhouse gases really are a huge threat"* is a weasel-word phrase that suggests that there is some validity in climate science but the agreement is not going to work. This echoes statements by collapse ideologues that surmise that there is nothing we can do – a defeatist attitude that underlies much of collapse ideology.

Trump's withdrawal from the Paris Agreement, and the positions promoted by Trump administration, have drawn high praise from collapse ideologues: Marc Morano wrote, "Trump's victory was seismic. It dramatically shifted the momentum away from policies based on the superstitious belief in catastrophic man-caused climate change and back to a rational energy policy—for now."[140] A *"rational energy policy"* for adherents of collapse ideology is promotion of coal, oil and gas, the three main contributors to global warming.

While lauding the Trump administration, Morano turns around and attacks those parts of the government that are actually doing climate work.

> With Trump's election, climate sanity was restored to the United States. No longer do we have to hear otherwise intelligent people in charge in D.C. blather on about how UN treaties or EPA regulations will control the Earth's temperature or storminess… Candidate Trump was the warmists' worst nightmare:

the first Republican presidential nominee who ever staked out a strongly science-supported skeptical position not only on climate change claims but also on the so-called "solutions."[141]

A *"science-supported skeptical position"* hardly describes the Trump climate policy. A more accurate description would be *"anti-science true believer position."*

State and City Actions

While the federal government has used dismissal of science to justify its anti-environmental actions, state and city governments have taken the opposite tack. A group of 17 states and numerous cities and counties, called the U.S. Climate Alliance, produced a statement:

> One need only look at the economies of our Alliance members to see that we account for 40 percent of U.S. GDP, at least $7 trillion dollars of combined economic activity, and 1.3 million clean energy jobs. Alliance states' economies have grown faster than the rest of the country, and we're continuing to grow as we make progress on our climate change goals…Together Alliance states have pledged to meet our share of the Paris Agreement

greenhouse gas reduction targets, to hold ourselves accountable by tracking progress toward those targets, and to accelerate the transition to a clean energy economy. We have made tremendous strides and the good news is that we are on track to meet or exceed our share of the Paris goals. [142]

While opposing the Trump Administration, the Alliance is organizing citizens, state and local governments to carry on U.S. policies in accord with the Paris Agreement: "We will do everything in our power to defend and continue our climate actions. This includes continuing to oppose any federal proposal to cancel the Clean Power Plan, weaken clean car and appliance standards or expand offshore drilling. One year after President Trump's abdication, the rapid economic growth of states within the U.S. Climate Alliance remain a beacon to all Americans and to every other nation that Americans are still in the Paris Agreement and will not retreat."[143] The Alliance has vowed to continue to honor the Paris Agreement and has sent representatives to Conference of Parties (COP) meetings in Bonn and Poland.

While states can continue the U.S. presence in international meetings on climate change, they cannot insure a coordinated policy for the entire country.

Some states – including Rhode Island and New York – have sued oil companies and some have sued the federal government, notably Massachusetts and California.[144] Cities also have sued oil companies for climate damage, but most of those cases have been dismissed. California has gone the furthest in implementing climate policy with its cap and trade instruments, but by itself that will not be enough.

Canada has recognized that working with U.S. states can be more fruitful than working with the federal government. Canada's Minister of Environment and Climate Change Catherine McKenna said "It is important to show the world that we're still working with U.S. states. There really are practical things we can do together."[145] Indeed, provinces such as Ontario and Quebec are cooperating with states such as California on measuring and limiting emissions. Foreign leaders are recognizing that the Trump Administration does not speak for the entire U.S. population.

States and cities in the Climate Alliance have been picking up some of the slack from the abandonment of the Paris Agreement by the federal government. They have done more to address climate policies than the feds. Now other states outside the Alliance, are fighting back, claiming that the Alliance states and cities are exceeding their legal authority and filing illegal lawsuits:

> "Federal courts should not use nuisance and trespass theories to confound state and federal political branches' legislative and administrative processes by establishing emissions policy (or, as is more likely, multiple conflicting emissions policies) on a piecemeal, ad hoc, case-by-case basis under the aegis of federal common law."[146]

Arguing that states should not use nuisance and trespass theories to fight climate change is like arguing that they should not use them for pollution or other threats to health. Legislative and emissions policies of the federal government are missing in action under the Trump administration, and it is incumbent on states to compensate for the lack of policies at the federal level. Alliance members extol the fact that "we are still in," at Conferences of the Parties serving as the Conferences of the Parties to the Paris Agreement.[147]

A cleavage in state actions is evidence of collapse ideology, where some states claim that climate policies of other states threaten their own well-being. States such as Montana, Wyoming, Colorado and the Dakotas have sued Washington State because Washington dared to challenge the need more fossil fuel facilities to ship coal, oil and gas overseas.

Some pundits have called the U.S. federal system a "laboratory of democracy" because states can diverge from federal policies and try out new approaches to problems such as climate change. This may be counterproductive in climate change policies. The question is whether state policies, even with states that represent 40% of the U.S. population, can compensate for the lack of action at the federal level. The State of Washington, for example, has 0.3% of world emissions, so reducing emissions to zero would affect only 0.3% of world emissions. It is necessary for all states in the U.S., and all countries around the world, to work together to implement the Paris Agreement.

Even progressive politicians in blue states such as Washington State tend to avoid the topic. In interviews with a public radio station, among the three progressive candidates in a contest for an open Congressional seat, one mentioned the issue only briefly.[148] Immigration and health care were the competing issues that got most of the air time, and while they are important issues their effects on society will be much less drastic that climate change. The Congressional district where the candidates are running includes parts of Washington State that are highly vulnerable to wildfires, which have been growing in intensity and size.

Washington State is also the locus of failed climate policy attempts, such as initiatives for carbon taxes that could not get more than 44% of the vote. Legislative actions have fallen short also, even though the legislature has put into law requirements for the state to reduce carbon pollution by 50% by 2050. There have been efforts to increase this percentage, which falls short of the Paris Agreement requirement of 100% reduction by 2050.[149] Even if it were increased, the state has yet to follow up with specific policies that would implement the limits.

Paris Agreement Problems

Even if leaders to adhere to the Paris Agreement, the issue of whether the agreement will do what is required to resolve the climate crisis is a live issue. Some authors have indicated that the Paris Agreement is a bad deal, but not for the reasons used by collapse ideologues – just the opposite. James Hansen, former director of the Goddard Institute for Space Studies and a notable leader of climate campaigns such as the Keystone XL Pipeline demonstrations, said "This [agreement] is half-assed and it's half-baked" because it would allow emissions to increase, albeit at a slower rate. He prefers a carbon tax rather than pledges to reduce emissions. Hansen's position is laudable but perhaps a little unrealistic. He amplified his criticism on

June 2018, the 30th anniversary of his widely-reported testimony at a Senate committee. "All we've done is agree there's a problem. We agreed that in 1992 [at the Earth summit in Rio] and re-agreed it again [at the 2015 Paris climate accord]. We haven't acknowledged what is required to solve it. Promises like Paris don't mean much, it's wishful thinking. It's a hoax that governments have played on us since the 1990s."[150] Hansen seems a bit overwrought when he calls it a hoax, but inaction on climate is a serious failure of the political system.

Some climate activists were even more derisive than Hansen. Friends of the Earth International, in a statement released on December 12, 2015, said "The climate deal to be agreed today is a sham. Rich countries have moved the goal posts so far that we are left with a sham of a denial." Naomi Klein said, "The deal unveiled to much fanfare and self-congratulation from politicians, echoed by an overly deferential press, will not be enough to keep us safe. In fact, it will be extraordinarily dangerous. We know, from doing the math and adding up the targets that the major economies have brought to Paris, that those targets lead us to a very dangerous future. They lead us to a future between 3 and 4 degrees Celsius warming."[151] Paris agreement goals (less than 2C with aspirations for 1.5C) have been reviewed by the IPCC (*see Appendix*), and to meet the goals the planet would have to rapidly implement some very "disruptive" energy policies. Most likely these are out of reach for most countries today, given the current political climate around the globe.

Collapse ideologues, from the opposite end of the spectrum, tend to dismiss the Paris agreement. Marc Morano said, "If we had to rely on the UN or EPA to save us from global warming, we would all be doomed. Nothing the UN is proposing to solve climate change would have any significant impact on temperatures or extreme weather events."[152]

More ominously, according to Morano, the UN has designs on the global capitalist system.

> The UN Intergovernmental Panel on Climate Change (IPCC) was formed in 1988 to examine how CO_2 and other greenhouse gases impact the climate. It had every incentive to declare a "crisis" because the UN was also going to be in charge of coming up with a "solution." If the UN failed to find CO_2 was a problem, it would also deny itself the opportunity to be in charge of regulating the world's economies and planning the energy mixes for the next hundred years and

beyond. This conflict of interest has been inherent in every action the UN climate panel has taken.[153]

It is absurd to contend that the UN is "going to be in charge of coming up with a solution;" the individual parties are responsible for solutions under the Paris Agreement, and they must decide what actions to take. IPCC reports provide options but they do not tell governments to adopt specific policies. (*See Appendix*)

Perhaps one of the strongest points about the Paris Agreement is that it allows for a variety of approaches and commitments to climate action. However, that could delay action until too late, when some of the "tipping points" of climate change come into effect. The agreement sets the limit for temperature increases at 2 degrees (C) and sets emissions limits at zero by 2050, but current commitments by the 197 parties to the agreement would only limit the temperature increase to 3.5C, and would not achieve zero emissions by 2050. Built into the agreement is a process of reviewing these commitments every five years and pressuring parties to increase their "ambitions," but there is no enforcement mechanism for achieving that. Some individual countries have pledged zero emissions by 2050: "Joao Pedro Matos Fernandes, Environment Minister of Portugal, said that Portugal is "committed" to becoming a carbon-neutral country by 2050 and is creating a national road map for this purpose, involving civil society, business and universities."[154] It will take more than a few small countries to achieve the goal of zero emissions by 2050 built into the agreement.

A separate, but related measure of success of the agreement, is carbon dioxide concentrations. Preindustrial CO_2 was at about 280 parts per million (ppm) concentration, and now we are above 410 ppm. James Hansen has indicated that a stable climate requires lowering that concentration to 350 ppm, and Bill McKibben named his climate movement 350.org. Scientists have warned that prospects for reaching this goal are not good: "the world's governments seem to be in no hurry to factor a 350-ppm target into their actions. To do so would require putting the earth on a crash carbon dioxide diet, a mass mobilization along the lines of World War II, to change the ways in which we all live, work, consume, and generate energy. Hansen also has advocated this approach for years. As politicians fiddle, however, evidence mounts that the climatic stakes are rising."[155] Not only would governments have to go on a "crash carbon dioxide diet," but also would have to scale up rapidly removal of CO_2 from the atmosphere, known as sequestration, to a much higher level just to get down from 410 to 350.

Petrodollars

What are the prospects for member states to increase ambitions and meet the Paris goals? Not good. One must examine the history of fossil fuel economies to understand the inertia built in to the international system. An excellent source for this examination is the book *"Carbon Democracy,"* an ironic title given the amount of antidemocratic activity that occurs in international petroleum politics and economics. One of the prime examples of inertia is the concept of "recycling petrodollars" through arms sales: "As petrodollars flowed increasingly to the Middle East, the sale of expensive weaponry provided a unique apparatus for recycling those dollars – one that could expand without any normal commercial constraint. … The financiers concerned with dollar recycling now had a powerful ally. Meanwhile, for the autocrats and military regimes of the Middle East, arms purchases provided a relatively effortless way to assert the technological prowess of the state."[156]

Recycling petrodollars is a particularly insidious way of managing the international economy. Wasting resources on military equipment and using up natural resources such as petroleum and the carbon capacity of the atmosphere are doubly destructive. In fact, the U.S. military realizes this in its own management practices as it attempts to reduce the use of fossil fuels in war zones, particularly Afghanistan. Some units use solar power because the cost of transporting fuel for use in vehicles and generators in war zones can be as much as $400 a gallon. Nevertheless, the foreign policy of the U.S. seldom considers the cost of climate change, particularly when it comes to dealing with oil states.

Undermining Democracy

A more ominous effect of climate change may be the undermining of democracy and the rise of authoritarian leaders as countries struggle with effects of climate change. An author of the International Transformational Resilience Coalition (ITRC) website, Bob Doppelt, describes this tendency as follows:

> Major shifts in the global climate could give rise to a new generation of authoritarian rulers, not just in poorer countries or those with weak democratic institutions, but in wealthy industrialized nations, too. Refugee crises, famine, drought – these are materials strongmen can use to build power. Already, strife and civil instability are spreading throughout the global South, with

droughts and floods stoking conflict and refugee crises in parts of Africa and the Middle East.[157]

As the effects of climate change tend toward causing collapse in social order, leaders will arise that promise to address these effects. Whether they are effective or not depends on whether they abandon the collapse ideology. If they do not, their heavy-handed methods may not succeed, and no one dictator can solve climate problems without international cooperation. International cooperation, however, is difficult for dictators to achieve, particularly when they seek to impose their rule on other countries. History teaches us that this kind of international order is not achievable.

Climate Abandoned

> *Along with God, nature is dying. 'Humanity' is killing both of them – and perhaps committing suicide in the bargain.*[158]
> Henri Lefebvre

Bill McKibben titled his 1988 book on climate change *The End of Nature.*[159] Perhaps humanity is abandoning both God and nature. In the process of abandoning nature, that is to say, isolating our activities from natural settings and ignoring the limitations that nature imposes on those activities, we are committing collective suicide.

We do not perceive our actions as suicidal, because we ignore the limitations set by nature on our use of energy and the atmosphere. But world leaders have begun to use the word *suicide:* UN Secretary General Antonio Guterres said, "Climate change is undeniable, the science is beyond doubt. It is time to get off the path of suicidal emissions."[160] While at the 2018 Conference of Parties of the UNFCCC, he said that failing to agree on implementation of the Paris Agreement "would compromise our last best chance to stop runaway climate change. It would not only be immoral, it would be suicidal."[161]

If his view becomes more widespread, action on climate change may be more likely, but so far Guterres is a voice in the wilderness. He is joined by a few others, such as Joanna Macy, who criticized the media for facilitating our tendencies toward collective suicide: "Corporate-controlled media seldom mention climate change, even when we're awash in record-breaking floods, hurricanes, droughts and firestorms. This institutionalized secrecy may protect vested interests, but it comes at a high price. Any system that consistently suppresses feedback – closing its perceptions to the results of its behavior – is committing suicide."[162] The responsibility of Corporate-controlled media for suppressing information about climate is also critiqued by some who dismiss climate science notably Rod Martin: "The corporate news media is owned by big corporations which, in turn, are owned by globalists who have their own, not-so-secret agenda. So many people do not trust big corporations, yet never give it a second thought to trust the evening news. And that's a problem. Climate always changes and always has."[163] Martin may be correct in asserting that we trust the evening news too much; the problem is not some secret agenda but the avoidance of climate news. Perhaps that is because it does not support the world view of the corporate media.

Reorienting Energy and Transportation

One reason that climate science dismissal persists and fuels inaction on climate is that people are afraid to face the implications of effective climate policy. Transportation in the U.S. accounts for a large part of U.S. emissions. Electric cars have been increasing in number, but at a smaller pace than internal combustion cars.[164] The recent addition of 1 million electric cars amounts to only 8% of the new cars added to the fleet. This means the U.S. is way behind the necessity to phase out gasoline and diesel powered vehicles. Reorienting the energy and transportation systems to zero carbon takes a major resolve by individuals, groups, businesses and governments to change how life activities are valued. Some have compared this effort to fighting World War II, when entire societies were mobilized for war production and military service. Consumer desires were set aside and everyone participated in the effort to fight fascism.

We are not making an all-out effort to address climate change, and we are delaying any efforts we could take way beyond their "use-by" date. Nathaniel Rich, in a long *New York Times* article, lays out the problem:

> The world has warmed more than one degree Celsius since the Industrial Revolution. The Paris climate agreement – the nonbinding, unenforceable and already unheeded treaty signed on Earth Day in 2016 – hoped to restrict warming to two degrees. The odds of succeeding, according to a recent study based on current emissions trends, are one in 20. [This is] a prescription for long-term disaster. *Long-term disaster is now the best-case scenario.*[165]

When long-term disaster is already "baked in" to our climate, Rich suggests that there are even more disastrous results if we cannot stop emissions soon. "Three-degree warming is a prescription for short-term disaster: forests in the Arctic and the loss of most coastal cities." The various kinds of disaster were outlined by Robert Watson, former director of the United Nations Intergovernmental Panel on Climate Change: "At four degrees: Europe in permanent drought; vast areas of China, India and Bangladesh claimed by desert; Polynesia swallowed by the sea; the Colorado River thinned to a trickle; the American Southwest largely uninhabitable. The prospect of a five-degree warming has prompted some of the world's leading climate scientists to warn of the end of human civilization."[166] Collapse is built into our current mode of fossil fuel use.

China, India, Brazil and a number of other vulnerable countries are members of the G20 (Group of 20), that includes the U.S., Canada, Australia, New Zealand, Japan and several members of the European Union as well as the EU secretariat. Together, these countries account for 85% of worldwide emissions, but they are doing little to slow them. In fact, the emissions have been growing after a slowdown from the lessened economic activity during the Great Recession:

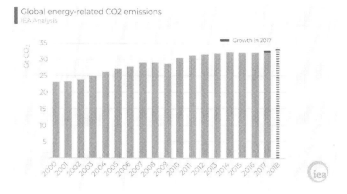

G20 countries contribute to increasing emissions not only from increased economic activity but also subsidies for fossil fuels. "It's in the G20's own economic interests to shift from brown to green energy, but we're still seeing major investment into the fossil fuel industry, along with huge subsidies," said Charlene Watson of the Overseas Development Institute (ODI).[167]

Unfortunately, with the withdrawal of the U.S. from the Paris Agreement, the G20 has become the G19. While the G19, minus the U.S., was issuing a declaration that supported climate action, the U.S. was touting fossil fuels.

- The G19 statement reads: "We will continue to tackle climate change, while promoting sustainable development and economic growth."
- The U.S. statement reads: "[The U.S.] reiterates its decision to withdraw from the Paris Agreement, and affirms its strong commitment to economic growth and energy access and security, utilizing all energy sources and technologies, while protecting the environment.[168]

It is surely an oxymoron to equate *"utilizing all energy sources and technologies"* with "protecting the environment." Not all energy sources will protect the environment; fossil fuels will do the opposite.

While waiting for the end of human civilization, some entrepreneurs are cashing in on climate change. After Hurricane Sandy, for example, some enterprising investors started thinking about protection of the New York City shoreline with massive projects. "A storm surge barrier system protecting New York City and parts of New Jersey could cost $2.7 million per meter," according to Michael Cembalest, JPMorgan Asset Management chairman of market and investment strategy. "In the early stages, people will be nervous. And the returns will be higher." [169] Investors may be more interested in adapting to climate change than mitigating it, as mitigation is a thorny problem involving government policy and opposition of powerful fossil fuels companies. "There is no way at this point to stop climate change. Pretty much every system is going to have to change. We're going to have to adapt to this" says James Everett, partner and co-founder at Ecosystem Integrity Management, a venture capital firm near Berkeley, California.[170] A fatalistic attitude like this may rationalize investment in adaptation for the short run, but it is not sustainable in the long run.

After Hurricane Harvey decimated Houston in 2017, some business people saw opportunities for cashing in. So many people lost homes that they were forced to stay in hotels, so investment in hotel properties became very profitable. "We saw occupancy go to 100 percent in a lot of those hotels. We didn't crush it. But we made 25 percent, 30 percent, pretty quick" said Rod Hinze, principal and portfolio manager at Key Point Capital in Dallas.[171]

Making money off climate change is a sign of collapse ideology when investors see little hope in mitigating climate change, but they do see major opportunities in adaptation. In effect, they are saying that "there's nothing we can do about it, so why not make some money." They are abandoning the idea that the current climate can be stabilized, and focusing on the likely opportunities of increased number and severity of disasters. Investment in renewable energy and reforestation might not be as profitable in the short run, but would pay much larger dividends in the long run. Focusing only on short-run adaptation is a form of collapse ideology that abandons climate stability.

We abandon nature and climate at our peril, using rationalizations that do not comport with science. If, for example, freedom and independence are translated into travel by single-occupancy vehicles fueled by gasoline and diesel, climate change will not accommodate our lifestyle no matter how we rationalize our behavior. If warm feelings about home and family require large separate households, houses will continue to be high-energy consumers. Manufacturers who view growth as increasing consumption of natural resources and use of the atmosphere as a free sewer system will worsen climate, whatever their motives.

Who is Responsible?

One of the most egregious attitudes about climate change is the feeling that there will be no responsibility of those in positions of power now. As Nathaniel Rich put it, the attitude was "Why take on an intractable problem that would not be detected until this generation of employees was safely retired?"[172] This attitude might have functioned unnoticed in the 1970's or 1980's, when political leaders could ignore climate scientists. John Sununu, former governor of New Hampshire and aide to President George H.W. Bush, described the indifference of political leaders at the time: "[A climate agreement] couldn't have happened because, frankly, the leaders in the world at that time were at a stage where they were all looking how to seem like they were supporting the policy without having to make hard commitments that would cost their nations serious resources. Frankly, that's about where we are today."[173] While politicians may still be uninvolved, it is no longer possible for the public to innocently ignore the signs of climate change.

All of these attitudes reflect the definition of insanity attributed to Einstein: "continuing to do the same thing over and over again and expecting different results." If we expect the climate to return to the stability it had during the rise of civilization for 10,000 years, while continuing to do what we are doing with fossil fuels today, we are fossil fools. We will have abandoned the climate to maintain our ideology of continued prosperity without changing our energy use.

Abandonment of climate change is possible only if people see themselves as separate from and independent of nature. Natural forces such as drought or changing temperatures were the causes of the demise of Easter Island and Greenland societies as described by Jared Diamond. In these small-scale cases, nature destroyed their unsustainable ways of life. Nature will always bat last. We are encouraged to abandon nature by our belief in a collapse ideology that encourages growth without regard to natural limits.

In today's world, we find a raging conflict between those who believe we can continue to grow our use of fossil fuels without limit, and those who recognize that there are natural limits that will increasingly impinge on our fossil-fueled growth. Without the recognition of natural limits, collapse ideology is shorn of reality and therefore encourages abandoning nature in favor of economic growth.

Climate Dystopias

While we may think of this folly as an example of humans abandoning nature, we will eventually find that nature abandons us. There have been a number of dystopic films on climate change, but two are particularly notable, one based on fiction and the other on non-fiction. *The Day After Tomorrow* is based on a premise of a sudden shift in climate after a freeze triggered by slowdown of the Gulf Stream. This is based on a hypothesis developed by some analyses of the effect of ice melts in the Arctic. Since land-based Arctic ice is mostly fresh water, it overlays salt water and causes it to change direction, reducing the warming effect of the Gulf Stream and stopping its movement to the north off Europe. The movie is sensationalist – the events it portrays could not happen in the short timeframe used by the script – but it has had a dramatic effect nevertheless and stimulated some discussion of climate change (including climate science dismissal by those who found the film preposterous).

A nonfiction documentary, *The Age of Stupid,* uses some news footage and other nonfiction ideas to portray a planet that had "gone south," deteriorating to the point where only a small area of the Arctic is habitable. The news footage comes from various events, including the denial of permits for a wind farm in the UK and the establishment of a new airline in India, to show how stupid the world has become. The last man alive narrates the film as he is sending these news items into space for inhabitants of other worlds to see.

More frivolous dystopias have portrayed humans as survivors in spaceships, such as *"Wall-e,"* the cute little robot that cleans up the earth left behind by irresponsible humans. The humans survive in a spaceship that caters to their every need while they lie around in floating lounge chairs. A more sober version is *Elysium,* a large donut-shaped space station that provides mini-gravity by rotating. It is furnished with greenscapes, lakes and nice houses for the inhabitants. Meanwhile on earth the environment is rapidly deteriorating and remaining residents are desperate to get to the space station.

Perhaps there are still serious thinkers that believe that we can abandon earth and survive, but it is more a fantasy than a reality. Clive Hamilton describes this fantasy: "Those who fly off leaving behind a ruined Earth would carry into space a fallen civilization. As the Earth receded into the all-consuming blackness, those who looked back on it would be the beings who had shirked their most primordial responsibility, beings corroded by nostalgia and survivor guilt."[174]

One thing that climate dystopias have in common is the concept that humankind will not survive, or will be drastically reduced in population by the effects of climate change. There is a kind of incredible collapse of civilizations built into this concept. Much is pure speculation, but the basis for doubting the future of humankind is present in climate change projections. While projections are built on models and subject to uncertainties, enough is known about the current effects of climate change to throw the future of humanity into doubt. One would think that this frightening prospect would cause widespread anxiety. And yet most humans are not aware, or do not comprehend, the seriousness of the threat to their own existence.

The Anthropocene

The irony of abandoning climate is that it will not abandon us. As some thinkers have put it, we have now entered into a new epoch, the Anthropocene, which differs from the previous epoch of the Holocene during which agriculture and civilizations developed. The Anthropocene is the era in which human activities for the first time will overwhelm natural balances in the atmosphere, biosphere and cryosphere, and will endanger our food supply and social stability.

Projections of changes in the Anthropocene depend on scientific models with ranges of uncertainty.[175] Uncertainty has another more deleterious effect on the discussion of climate change. While scientists are not 100% certain of any of the more extreme consequences of climate change, they are also uncertain of the milder outcomes of climate models. Michael Mann suggests that things could be worse than sometimes projected: "Just as climate models almost certainly underestimate the impact climate change has already had on such weather extremes, projections from these models also likely underestimate future increases in these types of events. Our study indicates that we can expect many more summers like 2018 – or worse." Mann, who has been the target of many climate skeptics who dislike his "hockey stick" analysis, notes that "Climate-change deniers love to point to scientific uncertainty as justification for inaction on climate. But uncertainty is a reason for even more concerted action. We already know that projections historically have been too optimistic about the rates of ice sheet collapse and sea-level rise. Now it appears they are also underestimating the odds of extreme weather as well. The consequences of doing nothing grow by the day. The time to act is now."[176]

Some people may still think that consequences of climate change are in the far future, but most have started to realize that they are with us now. Professor Kim Cobb has observed, "Decades ago when the science on the climate issue was first accumulating, the impacts could be seen as an issue for others, future generations or perhaps communities already struggling. In our increasingly muggy and smoky discomfort, it's now rote science to pinpoint how heat-trapping gases have cranked up the risks. It's a shift we all are living together."[177] We are now living in an era of change, and there will be no "new normal;" as we go further and further into the Anthropocene, "normal" will shift toward "extreme."

The Anthropocene will not be as congenial a period as was the Holocene. Clive Hamilton has projected what kinds of adjustment that we will need: "It is the narrative ... ventured in this book, of life lived and ordered under the shadow of the new geological epoch, in which the local is increasingly dominated by the global, where the events of history are already showing the stamp of global environmental change."[178] While in the past we could insulate ourselves from global events and forces through a kind of "fortress America" mentality, climate change will make that impossible soon.

It will be a new experience for humans to live under the new "geological epoch," for which many did not plan but were forced by their forbears to experience. By abandoning climate through their negligence of fossil fuel constraints, humans will be forced to deal with the new global forces of nature that will not spare anyone.

As Hamilton indicates in the above quote, we will have a "new narrative" in the Anthropocene, one that deals with the shadow of climate change. While collapse ideology has relied on the old narrative of human progress regardless of natural constraints, the new narrative will develop from the realization that abandoning climate is unsustainable. This realization will develop out of myriad small and large experiences with changes in the weather, as temperatures rise, the length of seasons changes, and extreme weather disasters become more frequent.

There is some resistance to attributing extreme weather to climate change, although the science of weather "attribution" is progressing. Nevertheless, there is increasing evidence of some climate causation for extreme weather such as droughts and floods – at least the strength of these events. In the case of tornadoes, there still is some uncertainty, but if we do reach 2 degrees warming, as mentioned in the Paris Agreement, "All the models agree that the frequency of tornadoes will increase by that point."[179] We will then have a new narrative about weather and climate: we will have to recognize causation by human folly.

No one will be free from this new narrative, although some may try to postpone its adoption as long as possible. Unfortunately, some of the narrative will be horrific as droughts and floods cause mass deaths, while other parts will be more hopeful as societies adapt and change. Some societies will lead the way and may provide support or models for others, but none can escape destructive effects.

Ideology and Collapse

> *The emissions produced by the cars running to and fro, meanwhile, will have their greatest impact on generations not yet born: they are so many invisible missiles aimed at the future...The person who harms others by burning fossil fuels cannot even potentially encounter his victims, because they do not yet exist. Living in the here and now, he reaps all the benefits from the combustion but few of the injuries, which will be suffered by people who are not around and cannot voice their opposition. Andrea Malm[180]*

So we have reached the final stage of analysis of climate science dismissal, the state of collapse so well described by Jared Diamond. As he described it, collapse results from a society's inability to recognize signs from nature and the environment of threats to its continuing existence. While applied mainly to small societies, Diamond's analysis also provides some hints for what we can expect on a worldwide basis from climate change.

Some authors such as Deb Ozarko consider our situation so dire that we are beyond the point of no return. "The reality is that time has run out. We are well past global warming, climate change and even anthropogenic climate disruption. We are in an irreversible state of runaway biosphere decay with no way back to what once was. And so I write this book with no hope-filled final chapter to lull anyone back to sleep."[181] Not everyone is so pessimistic about our future, but this viewpoint has begun to catch attention. There are signs that we are in a period of major climate disruption.

Changes in Weather

First to appear are the changes in environmental factors such as weather. While weather is short-term and climate is long-term, the frequency and intensity of weather events can signal impacts of climate.[182] The effects of climate change are destructive to the economy, as for example in the increase in extreme weather and wildfires. Scientists have estimated that climate change made Hurricane Harvey 40% stronger and increased the rainfall that flooded many Houston homes.

In the U.S. West, wildfires are increasing in intensity and frequency, and there is no longer a "fire season" in the summer and fall. Wildfires occur year-around and have become much more destructive.

These tendencies have been confirmed by scientists. "Scientists say that severe wildfire seasons in California may occur more frequently because of climate change. Since the 1970s, temperatures have risen by two degrees Fahrenheit in the western United States. And trees and grasses – the fuel of wildfires – are more dry and for longer periods of the year. That means fire season now lasts three months longer than it used to, starting earlier and often going on through the fall," said Jennifer Balch, a fire ecologist at the University of Colorado at Boulder. When extreme weather events could lead to breakdowns in infrastructure, the social structure of the planet will also deteriorate. First to go will be international organizations, including the UN and its constituent agencies, multinational businesses and multilateral nonprofits. Within countries, regions may break apart as communications and transportation systems are destroyed.

The future may resemble that portrayed in the movie "The Postman" where Kevin Costner transforms from a lone ranger to an impostor postman who links communities together with snail mail. Starting from a dystopia portrayed in the film, Costner's character gradually knits together fragmented regions and the U.S. becomes a whole country again. Unfortunately, this Hollywood ending is not likely to match our future climate prospects.

Conflicts and Refugees

More likely are regional conflicts with refugees such as those in the Middle East. The UN Security Council is charged with resolving regional conflicts, but it has only recently focused on climate change as a cause of conflicts. When the topic is raised, the discussion does not really address resolution of climate-based conflict. "The bar is still set low" Swedish Foreign Minister Margot Wallström acknowledged at a press conference, and said it was 'not realistic' to expect any immediate concrete outcomes, and that 'it's a success to be able to place it on the agenda.' "With more resource- and climate-related conflict likely on the horizon, it's difficult to know whether the Security Council taking on climate change is a hopeful sign, or just another indication that the world is marching steadily toward a future in which climate change poses a dangerous and unavoidable security threat."[183] In other words, the Security Council does not have the authority or resources to enforce standards such as those in the Paris Agreement. Should there be a peacekeeping force for preventing conflicts from climate change? Even asking this question arouses the fury of those who would reject any suggestion of a "world government."

Climate Scenarios

A number of scientists have described some likely scenarios that will result from climate change. One, Kate Marvel of Columbia University and the NASA Goddard Institute, summarized them as follows: "The future will bring upheaval and uncertainty. Sometimes disaster will be imprinted with the undeniable fingerprint of climate change. Glaciers melt, oceans warm, and rising seas will swallow small islands and coastal cities. Their residents will never be able to return." Island residents may be the first climate refugees with no homeland to which they can return. "These people will have a clear and compelling claim to asylum; they will flee total destruction clearly attributable to a warming climate. The wealthy countries that have emitted the vast majority of greenhouse gases will bear moral, if not necessarily legal, responsibility for their plight." This is a prime example of climate injustice, described previously. "But there will be other categories of refugee in the coming world: farmers struggling to grow profitable crops in drying soil, manual laborers whose working hours are curtailed by heat and humidity, despised minority groups conveniently blamed for new adversities." If climate change is, as our military considers it, a "threat multiplier," there is no shortage of existing threats to multiply. The Earth does not warm independently of the people who live on it.[184]

Climate refugees will become an increasingly severe problem, as Europeans are discovering. The EU is being split apart by differing views on the refugees from the Middle East and Africa, many of whom are fleeing drought conditions in their home countries. Who bears moral responsibility is a profound question. While advanced countries have resources to adapt to climate change, they also have the means to aid less advanced countries with their energy systems and their droughts and floods. Yet this issue of finance has plagued negotiations on climate change for many years,[185] and the U.S. Senate in 1997 voted 95-0 against the Kyoto Protocol because less developed countries were not required to reduce emissions. That has been rectified in the Paris Agreement but the Senate has not changed its attitude. The U.S. federal government, as noted in previous chapters, has dedicated its climate policy to opposition to international cooperation on climate policy. This has isolated the U.S. from the rest of the world. "You have this enormous discrepancy between the White House and, essentially, everyone else. The leadership in Washington is really moving against the whole agenda," said Johan Rockström, the director of the Potsdam

Institute for Climate Impact Research in Germany.[186]

Federal Government Abandons Policy

Federalism, coupled with avoidance at the national level, means that the states are bearing a major burden of U.S. climate policy. Already U.S. states are diverging sharply on their approaches to climate change. Northeast and West Coast states are addressing climate change directly while much of the rest of the country is passive or actively opposed to climate policies. Ironically, one of the most positive developments is the rapid increase in wind generation in Texas and other mid-continent states. While their governments continue to act as if fossil fuels will always be the energy sources of choice, businesses and consumers are shifting away from fossil fuels.

Not only in the U.S. have businesses, cities and states started to take climate action. The UN Environment Program reports, "The number of actors participating is rising fast: more than 7,000 cities from 133 countries and 245 regions from 42 countries, along with more than 6,000 companies with at least US$36 trillion in revenue, have pledged mitigation action. Commitments cover large parts of the economy and are gradually expanding in regional coverage. Many of the actors are engaging in so-called 'international cooperative initiatives', which are characterized by multi-country and multi-actor engagement."[187]

Even though some climate actions are underway at sub-national levels, it is not enough to do what some states such as California, Washington State and the Northeast have started. In Washington State, for example, the current goal is 50% reduction of emissions by 2050 while the Paris Agreement mandates a 100% reduction. Even 50% is more than most states will commit. Ratcheting up to 100% by 2050 is progressively more costly with each year of delay, but failing to reach this goal condemns the country to collapse. A failed Washington State initiative would impose a carbon fee of $15 a ton on carbon emissions. But the IPCC report estimated that governments would need to impose effective carbon prices of $135 to $5,500 per ton of carbon dioxide pollution by 2030 to keep overall global warming below 1.5 degrees Celsius, or 2.7 degrees Fahrenheit.[188] Even at $15 a ton, the Washington State proposal drew a lot of opposition so one could imagine how much opposition $135 to $5,500 per ton would draw.

As federal and state governments fail to act, some citizens are taking on the legal system to try to spur action. Often these efforts take the form of lawsuits on behalf of young people, whose future is most in peril from lack of climate action. Lawsuits at the state level in Washington State, California and other states have made their way through the courts, often opposed by state and federal governments on the grounds that policy should be decided by legislative action, not the courts. Of course, legislatures have not acted in most states, nor has Congress at the federal level, and the Trump administration is rolling back Obama-era regulations. Nevertheless, legal groups such as Our Children's Trust have pursued lawsuits and have been granted standing, at least at the district court level.[189] It remains to be seen if these lawsuits will survive appeal.

States and cities have also filed lawsuits, including some in California where a judge agreed to hear arguments but eventually denied standing to the plaintiffs. In his opinion, he made the argument that other government entities are more responsible for climate policy than the courts:

> This order fully accepts the vast scientific consensus that the combustion of fossil fuels has materially increased atmospheric carbon dioxide levels, which in turn has increased the median temperature of the planet and accelerated sea level rise. But

questions of how to appropriately balance these worldwide negatives against the worldwide positives of the energy itself, and of how to allocate the pluses and minuses among the nations of the world, demand the expertise of our environmental agencies, our diplomats, our Executive, and at least the Senate.[190]

As Stalin said, the pope has no regiments; we can say that the courts have no police powers to back up decisions on climate change. We depend on legislatures and executive agencies to carry out climate policies and for the most part they have been found wanting. Courts do have a role to play, however, when other branches of government have made climate policies and been challenged in court. In Washington State, for example, a shipping company was denied permits for a coal terminal, partly on the basis of high emissions from burning coal. Seven states, including Wyoming, Montana and Colorado, sued the state of Washington. This case is still pending as of this writing. If the courts decide in favor of Washington State, it will be a small victory for those urging state actions.

Will They Follow Our Example?

Other countries vary in their plans for reducing emissions. With the lack of

U.S. action at the federal level, other countries may divide into two camps: (1) those who will go ahead with their plans regardless of the U.S. policies, and (2) those who may ease or delay their actions and follow the example of the U.S. Those in the first camp who meet their Paris commitments and ratchet up their efforts to reduce emissions can lead the way but that may not be enough. Europe is most likely to be in this camp and continue to strengthen its efforts; China and India may follow Europe's example but have a long way to go. There are question marks about other countries such as Canada, Japan and Russia who have backslid from some of their commitments (e.g., they quit the Kyoto Protocol). Countries that follow the U.S. lead, which may include Australia[191], Canada,[192] Brazil[193] and others, are likely to find that their lack of action makes them pariahs as climate consequences become worse. The UN Environment Program has reported that "Argentina, Australia, Canada, the European Union, Indonesia, Japan, Mexico, South Africa, the Republic of Korea and the United States are likely to require further action in order to meet their NDCs, according to government and independent estimates."[194] While it is diplomatic language to say that these countries are "likely to require further action," the UN is really scolding them.

There is some hope that shame might have a positive effect. That is a faint hope, however.

Trump's withdrawal from the Paris Agreement also exacerbated relations with China, with whom President Obama had worked to make the Paris Agreement come to fruition. China and the U.S. are number one and number two, respectively, in the highest greenhouse gas emissions. "The biggest threats to the planet are the lack of U.S. climate leadership at home and the unwillingness of the U.S. to engage with China. The rest of the world looks to the U.S. and China for leadership, and it has become clear that, as the alliance has waned, global momentum to address climate change has slowed," according to Joanna Lewis, a China specialist at Georgetown University.[195] Slow momentum in addressing climate change is the last thing the world needs now.

Signs of backsliding by countries began to emerge in 2018 negotiations on the follow-up to the Paris Agreement. "Recent negotiations in Bangkok on the Paris Agreement's implementation guidelines made some progress, but not enough. Nations are not living up to what they promised," according to Patricia Espinosa, the executive secretary of UN Framework Convention on Climate Change.[196] As has been noted, the Paris Agreement is itself insufficient for addressing the climate crisis and any backsliding would exacerbate global failure to address climate change.

A major factor in the backsliding of some countries is the "Trump Effect." As described by Joseph Curtin, senior fellow at the IIEA and the author of the report: "The 'Trump effect' has created a powerful countervailing force acting against the momentum (the Paris Agreement) hoped to generate." Curtin described three key areas in which the Trump effect had "applied a brake" to the global climate actions: "(1) federal rollbacks were increasing the attractiveness of fossil fuel investments; (2) U.S. withdrawal from the agreement had created moral and political cover for other nations to follow suit; and (3) goodwill at international negotiations was being damaged."[197] Morality is not playing a major role in U.S. foreign policy, and to expect others to maintain a moral stance while the U.S. abandons it is not realistic.

During the December 2018 Conference of Parties of the UN Framework Convention on Climate Change in Poland, the role of the U.S. as a dissenter (discussed above) led observers to note its deleterious effect on global unity. "The fact that nations are spending this much time on minor wording issues while the science finds increasing risk of catastrophe has to be seen as a metaphor for how inadequate the global response to the climate challenge has been," said Paul Bledsoe, a former Clinton climate adviser. "It also shows that the lack of U.S. leadership has massive costs to global ambition."[198] These costs include not only the climate change effects exacerbated by U.S. emissions, but the knock-on effect of U.S. backsliding that enables other countries to indulge in collapse ideology.

Without effective global action, the world is headed for collapse. "End of the World" themes have most often been associated with apocalyptic religious cults, who rely on the Book of Revelations in the Bible or other sacred writings to predict collapse. Now, however, there are philosophers of science, like Timothy Morton, making the same prognostications:

> The end of the world has already occurred. We can be uncannily precise about the date on which the world ended. It was April 1784, when James Watt patented the steam engine, an act that commenced the depositing of carbon in Earth's crust – namely, the inception of humanity as a geophysical force on a planetary scale.[199]

Of course, the proceeding 235 years have not seen a complete collapse, and we will not likely see it in our lifetimes. But the insight in this statement is that the use of fossil fuels has changed societies in ways that will be next to impossible to reverse.

Apocalypse Now?

Evangelical Christians tend to adopt an apocalyptic view of the future, but there are indications that they use theological reasons to avoid concern about climate change. A 2015 Pew Research Poll, the most recent on the subject, found "only 28 percent of white evangelicals believed that the Earth was getting warmer because of human activity – by far the lowest percentage of any religious demographic in the survey."[200] They may have heartfelt beliefs about nature and the divine that permit this dismissal of climate science, but the effect is to impede the adoption of effective climate policies. In other words, there is a kind of willful ignorance of science when it does not fit the theology. This form of ideology is more of a passive denial, since the concept of collapse does not comport with their beliefs.

Others do anticipate an apocalypse. When the California state government issued a report, the *2018 Climate Change Assessment,* Governor Jerry Brown said, "These findings are profoundly serious and will continue to guide us as we confront the apocalyptic threat of irreversible climate change."[201] California has been especially hard hit by climate change, including droughts, wildfires and coastline erosion from sea level rise.

On a more positive note, some thinkers have come up with enterprising solutions to climate consequences such as sea-level rise. A company in Holland, Dutch Docklands, has entered into a joint venture with the Maldivian government to build "floating megaprojects," including "a floating golf course complete with undersea transparent tunnels between eighteen holes offering aquarium views of wild tropical fish and manta rays, and Greenstar, a floating grass-terraced hotel in the shape of a starfish, where each arm of the starfish is removable and replaceable, like a cruise ship. These projects will total 80 million square feet of floating space. Currently for presale are 185 floating villas to be arranged in the shape of a flower."[202] These all sound very fanciful. Nevertheless, the Maldives and Bangladesh are like canaries in the coal mine, and their survival may depend on innovations such as seasteading. As agreeable as this seems, it is depressing to think that an island nation, or a large nation with much of its land on a river delta, will be submerged and depend on artificial land to stay afloat.

Anticipating collapse is a major psychological problem. People can become numb when confronted with the possibility of collapse. Robert Lifton, who has studied the effects of nuclear war (including victims in Hiroshima) has made this observation: "Everyone calls forth a measure of psychic numbing toward nuclear and climate threats. Such numbing, as I observed in earlier work on the effects of nuclear weapons, has to do with the mind's resistance to the unmanageable extremity of the catastrophe, to the infinite reaches of death and pain."[203] This numbness affects even the activists who are fighting against the growing use of fossil fuels. They know the effects of fossil fuel usage, but while working on climate solutions, they may find themselves using fossil fuels for transportation and sustaining their lifestyle even though they realize that this use contributes to the problem they are trying to solve. It also opens some like Al Gore to criticism from those who are skeptical of climate action.

Such a paradox is evidence of the nature of climate change as a "wicked problem." While we try to understand what we are doing and act to prevent further damage, we still find ourselves dependent on the very thing that we detest.

One indicator of the persistence of dependence on fossil fuels is the growing development of unconventional sources of crude oil. Fracking and the carbon-intensive extraction of oil from tar sands are prime examples of this. Wainwright and Mann have noted that "…this shift signals the end of any hope for meaningful carbon mitigation. Fracking and related extractive processes are much more carbon-intensive than drilling Saudi oil, and the explosion in unconventional hydrocarbons guarantees increased greenhouse gas emissions." They relate climate consequences to social consequences such as income inequality: "In addition, the geographic and political-economic distribution of these resources deepens the global division of wealth and power, exacerbating geopolitical inequalities and further destabilizing what little ground international negotiations have cleared for cooperation on climate-related concerns."[204]

International Cooperation

Even efforts at achieving international cooperation on climate have fallen short. In Copenhagen in 2009, President Obama met with counterparts from China, South Africa, Brazil and India to hammer out a joint communique, but it was not enough for a ratified agreement and also fell short of what was needed. Mary Robinson, former president of Ireland, noted that "Although Copenhagen produced the first joint commitments on emissions by major developed economics, these cuts fell far short of what [Kiribati President] Tong and leaders of other vulnerable countries had been holding out for in hopes of keeping the global temperature rise to 1.5C in this century."[205] The special IPCC report (see

Appendix) calls for a complete disruption in the current fossil fuel political economy in order to stay below that 1.5C level. But the current political economy continues free of any disruptions.

Unconventional oil extraction in the U.S., Europe, China and Canada has empowered them to continue high levels of fossil fuel use without adjusting to any changes in world order that might reduce their wealth and influence. Despite all of the international action aimed at reducing emissions, embodied in the Paris Agreement, these countries can thwart international cooperation and render the agreement meaningless, as the Trump decision has revealed.

Oil Exploration

Absurd exploration for new sources of oil using unconventional techniques is a sign of fossil-fuel ideology leading to collapse. When scientists estimate that four-fifths of the *known reserves* should remain in the ground, to develop new reserves is the height of folly. Pope Francis told oil company executives that it is wrong to continue the *"search for new fossil fuel reserves, whereas the Paris agreement clearly urged keeping most fossil fuels underground".*[206] Not only is continuing oil exploration a threat to the world's climate, it is financially irresponsible if oil companies may be acquiring stranded assets, i.e. reserves that would not be usable *if* climate action becomes effective. That, of course, is a big "if." It depends on abandoning collapse ideology, but societies continue to adhere to the ideology to avoid facing reality, at their peril.

An example of the absurdity of oil exploration is Alaska's National Wildlife Refuge. While costs of exploration and production are higher than other areas, the politics of Alaska plus the "dividend" received by all Alaska residents make expanding production into ANWR desirable. Senator Lisa Murkowski promoted it for years, even though President Bill Clinton vetoed an exploration bill and administrations of Presidents George W. Bush and Barack Obama did not allow drilling. But when President Donald Trump was elected, Murkowski was able to get drilling added to a tax cut bill on the pretext that it was a revenue-raising measure, not an environmental issue.[207]

One reason that this absurdity remains in place is that political systems depend on assumptions of growth in the use of fossil fuels. In fact, one author, Timothy Mitchell, argues that democratic political systems depend on this assumption: "Democratic politics developed, thanks to oil, with a peculiar orientation towards the future: the future was a limitless horizon of growth."[208] There is a dark implication in this statement: as natural limits to the use of fossil fuels become apparent with the effects of climate change, democratic political systems will fail. As floods, droughts, heat waves and other effects of climate change become more pronounced, voters may support dictatorial candidates who promise relief. China offers a model for this kind of political reaction to climate change. With a command economy that can switch rapidly to electrification of transportation and the use of renewable energies on a vast scale, China seems to lend credence to the idea that an authoritarian system is the only recourse for failing Western economies. China's experiment is in its early stages and still has a lot of fossil-fuel legacy to overcome. Most of the world is still saddled with fossil-fuel habits that mean decades of emissions that will lead to CO_2 concentrations at an irreversible level.

It is this irreversibility that makes collapse likely. Carbon dioxide remains in the atmosphere for centuries and will continue to trap heat long after emissions have ceased. There are some who think geoengineering will be necessary to avoid collapse, but the prospects for geoengineering actions without unintended consequences are limited. Only a few, such as reforestation and carbon-capture-and-storage, are not risky; but scaling them up to the level needed to avert collapse is enormously costly.

Ideology Persists

If ideology is associated with collapse, why do people continue to believe in the ideology? Malm offers a suggestion: "The referent of 'ideology' has undergone a slippage, from a system of ideas proclaimed by meetings and monuments to a structure so deeply ingrained in the very materiality of ... society as to be invisible, inaudible, crushingly efficient because it is unstated and taken for granted."[209] This definition of ideology gives us some tools for analyzing why people continue to deny, explicitly or implicitly, climate science and the consequences of collapse. Ideology is not always consciously understood or admitted, but it can nevertheless be *"crushingly efficient because it is unstated and taken for granted."* The efficiency of ideology to which Malm refers is its ability to keep people on track to collapse despite their own awareness that change is needed to avoid collapse.

There is a kind of willful ignorance in collapse ideology, a tendency to downplay or dismiss climate science when it does not permit continued use of fossil fuels. Clive Hamilton has described this well: "Our best scientists tell us insistently that a calamity is unfolding, that the life-support systems of the Earth are being damaged in ways that threaten our survival." Even so, we still indulge in false beliefs: "Yet in the face of these facts we carry on as usual. Most citizens ignore or downplay the warnings; many of our intellectuals indulge in wishful thinking; and some influential voices declare that nothing at all is happening, that the scientists are deceiving us."[210]

It is perhaps more comforting to ignore the signs of climate change than face them. This is a function of ideology, which can be defined as fixed beliefs about a changing reality. As reality intrudes on our fossil fuel lifestyle, we prefer to ignore it rather than face it. The concept of "cognitive dissonance" helps explain this tendency to dismiss reality rather than change our beliefs.

Ideology plays a major role in collapse. As human civilization develops fossil fuel infrastructure and becomes more and more dependent on it, there is little motivation to change and strong motivation to justify continuing on the same path. This motivation leads to avoidance of necessary changes through a kind of ideological inertia – belief that there is no need to change. This has been well described by an author Physicist Adam Frank: "If you change the earth's climate enough, you might not be able to change it back. Even if you backed off and started to use solar or other less impactful resources, it could be too late, because the planet has already been changing." He then coins a phrase for the necessary changes identified by climate science: "These models show we can't just think about a population evolving on its own. We have to think about our planets and civilizations co-evolving."[211]

"We have to think about our planets and civilizations co-evolving" is a statement of climate action that is key to the understanding collapse ideology. Any way of thinking that separates civilization and nature leads to collapse. Economic theories that posit infinite growth, or any growth that ignores natural limits, are an example. Beliefs in maintaining a lifestyle independent of natural constraints are another example. While these are not explicit ideologies with a full set of climate denial tenets, they are a kind of passive denial ideology that leads to the same effects.

The Cost of Carbon

One of the crucial concepts in climate action is the cost of carbon. When calculating the cost to health and the environment, governments can assign an estimated cost to each ton of emissions. In the Obama Administration, the cost of carbon was set at $50 a metric ton. This was low, perhaps way too low, but now the Trump Administration is putting it much lower – at from $1 to $7 a ton.[212] There are two problems with this low estimate: the cost is calculated only for damages in the U.S., and only for the current generation.

Such a narrow calculation illustrates two aspects of collapse ideology: climate change is viewed as a national environmental issue, and its effects are considered minimal for the near future. Both are false. Carbon emissions and concentrations do not respect national boundaries. Carbon emitted anywhere in the world can affect all other locations because the concentrations spread worldwide and have effects everywhere, regardless of where they are emitted. Carbon accumulates in the atmosphere, and can continue to heat the planet for centuries even if emissions were reduced drastically.

Estimates of the worldwide cost of carbon are as high as $417 a ton, way above either the Obama or Trump administrations' figures.[213] This is because the costs accumulate as the carbon accumulates in the atmosphere, and some countries are impacted much more than others.[214] Estimates are, of course, only projections but many projections about climate change have been lower than current data have already demonstrated.

Another aspect of the cost of carbon that makes climate change such a "wicked" problem is that resolving the effects of climate change becomes more costly each year. Reducing emissions thirty years ago, when climate change became widely known to politicians as well as scientists, would have been relatively cheap. It is now much more expensive. Not only to emissions have to be reduced at a much faster rate to avoid collapse, extensive sequestration must be initiated. Each of these processes is much costlier now than thirty years ago.

In the immediate future, calculation of the costs of carbon is likely to be minimized for political expediency, but this cannot last. Heat waves, floods and wildfires become more costly each year and the costs cannot be reduced by ignoring climate science; they will only increase.

Who pays the cost of carbon? At present the public pays in many different ways, including health costs and disaster recovery costs. Payment of these costs is *ad hoc,* highly irregular and difficult to calculate because it involves so many different costs imposed on victims on a somewhat random basis.

Financial analysts estimate that paying the ultimate cost of carbon will shock the financial system, and will include the wealthy as well as the general public.

Investment firm Schroders said "there could be $23 trillion of global economic losses a year in the long term without rapid action. This permanent economic damage would be almost four times the scale of the impact of the 2008 global financial crisis." Standard and Poor's rating agency also warned leaders: "Climate change has already started to alter the functioning of our world."[215] When financial analysts start calculating these enormous sums, we know that carbon costs will be severe and unavoidable.

Economists have calculated costs of carbon and devised much more systematic ways of assessing them. One of them, Yale's William D. Nordhaus, received a share of the 2018 Nobel Memorial Prize in Economic Science for his analysis. He suggested that "the most efficient remedy for the problems caused by greenhouse gas emissions would be a global scheme of carbon taxes that are uniformly imposed on all countries."[216] This makes sense in an eminently rational way, but it runs head-on to practical politics. Taxes are viewed adversely on a worldwide basis, and the concept that they could be "uniformly imposed on all countries" would inflame the passions of many believers in the paramount sovereignty of nation states.

Discussion of carbon taxes reminds us of the reason that climate change is considered such a "wicked" problem. Costs of climate change are mounting but they are not fully paid by the people whose activities are the source of the problem. Any attempt to right this injustice is met with fierce resistance, and nothing gets done until the problem becomes so bad that we are desperate.

President Donald Trump, who had once called climate change a "hoax," has changed his tune and now admits that the climate is changing, but extends the time horizon: "I'm not denying climate change. But it could very well go back. You know, we're talking about over a ... millions of years."[217] Trump is correct in that climate will of course change over millions of years, but we are more concerned about the next decades.

As with many collapse ideologues, Trump combines condemning the cost of climate action with ignorance of the cost of carbon. "But I don't know that it's manmade. I will say this: I don't want to give trillions and trillions of dollars. I don't want to lose millions and millions of jobs."[218] No one would call Trump stupid, given his ability to elicit support from Russians, and to question human causation when he knows better is a sure sign of collapse ideology. When he couples feigned ignorance with a one-sided analysis of carbon costs, he provides a prime example of collapse ideas. The true costs of carbon will far outstrip the millions of jobs and trillions of dollars that Trump assigns to climate action.

Trump's attitude reflects a major feature of collapse ideology: refusal to ascribe costs to carbon emissions, while attributing enormous costs of doing anything about them. This attitude is expressed in an exaggerated way by another apostle of dismissal, Norman Rogers:

> The global warming believers claim all sorts of fantastic disasters from CO2 emissions, so any cost can easily be justified for reducing CO2. The money is being spent now for a future, highly dubious, and highly discounted event. My opinion is that the well-known benefits of CO2 for agriculture, making plants grow better and with less water, outweigh any benefit from reducing CO2 emissions. Thus, there is every reason to welcome CO2 emissions.[219]

Is CO2 Beneficial?

One peculiar feature of collapse ideology is apparent here: increased CO2 can be beneficial because it enhances plant growth. Plants are sensitive to things other than CO2 in the environment. In particular, high temperatures reduce crop yields and drought often stops growth altogether.

Roger's designation of climate activists as proponents of the argument, that "any cost can easily be justified for reducing CO2," is a backwards way of analyzing costs of carbon. It is not the climate activists that justify costs for reducing carbon; it is the costs themselves that justify climate action. Ignoring the already documented costs and the huge projected costs is a sign of collapse ideology. Here are some of the already documented costs:

- On average, floods and storms have displaced nearly 21 million people every year over the last decade, according to the Internal Displacement Monitoring Center. That is three times the number displaced by conflict
- Worldwide, according to Munich Re, damaging floods and storms have more than tripled in number since the early 1980's. These economic losses have risen sharply with two record years in the last decade in which damages topped $340 billion.[220]

In other words, carbon now costs $340 billion a year, twice the average of $170 billion a year for the prior ten years.[221]

Another way of looking projected carbon costs is to see how many people, species and habitats will be affected by increased temperatures, measured as the difference between 1.5C and 2.0 C. The IPCC report (Appendix) provides many examples of increased costs that will impinge on the environment, as temperatures rise above 1.5C even if we stop the rise at 2C.[222]

How are corporations managing carbon costs? After all, corporate business managers are supposedly experts at accounting for costs. Indeed, many are building carbon cost equations into their financial statements, and regulators are starting to insist that they do so. Nevertheless, there is a strong temptation to avoid accounting for costs. Exxon-Mobil, which for years denied climate science (at least in public), was sued by New York State because it underestimated its climate vulnerability:

> New York's lawsuit paints a picture of corporate malfeasance whereby Exxon told investors it was applying a "proxy cost" to its extraction of fossil fuels to account for the risk of governmental action to address climate change. In practice, the lawsuit alleges, Exxon applied much lower proxy costs than it initially stated, essentially assuming that existing climate regulations would remain in place, unchanged.[223]

It may be a comforting assumption that *existing climate regulations would remain in place, unchanged*, but that is a major flaw introduced by collapse ideology. Costs will increase regardless of government actions; either the price of carbon increases first or the consequences of climate change impose major costs later. Corporations such as Exxon-Mobil will not escape those costs.

Geoengineering

The concept of geoengineering arose out of desperate thinking about what to do with climate change when the consequences become intolerable. Various schemes with widely ranging risks have been proposed, including cooling the earth with reflective aerosols, seeding the oceans with iron to encourage plankton uptake of CO_2, and sequestering carbon on a vast scale. Ironically, some of these are promoted by collapse ideologues as a way to avoid collapse: Katherine Ellison notes, "Amid these developments, some close allies of Trump have taken a seemingly paradoxical stance: While denying climate change is a human-caused problem and rejecting proposals to cut greenhouse gases, they're promoting what many experts worry is the risky default solution of geoengineering."[224]

Desperation is the key to understanding geoengineering. Many climate scientists, including James Hansen and Paul Crutzen, have despaired at the inability of humanity to resolve climate disruption before it is too late and feedback cycles make it unmanageable. As Johansen has warned, "A tone of desperation is palpable in climate-change science when well-known people seriously propose that filling the stratosphere with sulfur dioxide may be the only way to stop runaway greenhouse warming."[225]

Unintended consequences from geoengineering such as changes to weather patterns and acid rain from sulfur dioxide make it questionable at best. Desperation is also a factor for those who currently deny climate science but may find it necessary to "catch up" when climate consequences become severe.

Some will continue to promote geoengineering, including conservative think tanks, because it is an alternative to actually dealing with the basic problem of carbon emissions. Clive Hamilton has observed that "Rather than slashing the asset value of some of the globe's biggest corporations, asking consumers to change their habits, or imposing unpopular taxes on petrol and coal, this form of solar geoengineering carries the implicit promise that it will protect the prevailing politico-economic system, which is why certain conservative American think tanks that for years have attacked climate science as fraudulent have endorsed geoengineering as a promising response to global warming."[226]

So it seems that while denying there is a problem, some geoengineering advocates propose a solution that is worse than the problem, to avoid confronting the real issue.

Plant growth seems to offer some ideologues a way to address the principle cause of climate change, the increasing concentration of carbon dioxide. Indeed, some collapse ideologues have attributed the "greening of the planet" to increased emissions of carbon dioxide, and have claimed that CO2 is a plant food, not a pollutant. They may cite the research of Elliott Campbell, an environmental scientist at the University of California, Santa Cruz, who found that plants are now converting 32 percent more carbon dioxide into lignin and other plant matter than before the Industrial Revolution.[227] Dr. Campbell cautions, however, that this research does not indicate that plants can "solve" climate change, as the increase in CO2 emissions is overwhelming the ability of plants to absorb CO2 and is increasing heat stress on plants. Nevertheless, some ideologues claim that plants will solve problems of climate change.

It seems ironic that those who deny climate science would propose solutions for a perceived "non-problem." But it is clear that while they deny the science, they are considering options for responses when the consequences become unavoidable. This always raises the question of motivation: why are they denying the science when they know it will determine their future? It is a sign of collapse ideology when proponents need to hedge against the future they deny will happen. Do they really believe that there are no consequences of climate change that would lead to collapse?

Future Climate Migration

We will see some of the first consequences of collapse in increasing human migration.[228] Already coastal areas of the U.S. are seeing increasing migration; after hurricanes, areas of New York, New Jersey, Virginia, North and South Carolina, Florida and Louisiana have become uninhabitable. Alaska natives find it necessary to move entire villages. "I don't see the slightest evidence that anyone is seriously thinking about what to do with the future climate refugee stream. It boggles the mind to see crowds of climate refugees arriving in town and looking for work and food," said Orrin Pilkey, professor emeritus of coastal geology at Duke University.[229]

Migration is a moral issue: the people of Bangladesh, for example, are not responsible for the sea level rise inundating their country. They are responsible for fewer carbon emissions than those from more industrialized countries. Entire Pacific island countries may disappear. The Unitarian

Universalist Service Committee (UUSC) has a project dedicated to helping island nations adapt.[230] The moral issue of climate displacement is described by UUSC:

> The threat of climate-forced displacement is disproportionately acute in small developing states and indigenous communities in remote areas. These communities are often under-resourced and politically marginalized, and, in some cases, have histories that include dealing with environmental change, tribal conflict, and earlier displacement by colonialist or corporate land grabs. Communities are often carrying out resettlements or relocations without legal protections and inadequate funding from private or governmental sources. Above all, global efforts to mitigate and adapt to climate change must be guided by human rights norms and principles, including the rights to participation, self-determination, transparency, and nondiscrimination.[231]

It is clear that climate change is much more than an inconvenience for some residents of coastal areas. As climate migration makes clear, there is a profound, life-altering effect of climate change on vulnerable communities. Eventually, climate change will impact everyone and the probabilities of human survival are declining each year.

Human Extinction

What if civilization collapses and humans disappear from earth? Some have said that may be inevitable, that all that will be left are insects and reptiles (and presumably some plants thriving on all the CO_2). Others have said that earth can shrug off the effects of human folly and continue, albeit with a different climate. Adam Frank has said that we should recognize that we are not destined to inhabit earth forever. "This recognition – that in the long term the Earth will abide without us – does not absolve us from the need for urgent action. It is not an excuse for climate denial or ecological hooliganism. It also does not mean we are free to just impose suffering on Earth's other creatures. Instead, it's an acknowledgment of the true scale of our planetary responsibilities." What are those responsibilities? "It means we must become the agent for something the Earth has not seen before — a biosphere that is also awake to itself and can act for its future with both compassion and wisdom.[232]

Perhaps we will recognize the true scale of our planetary responsibilities, but that will require that we overcome the tendency toward collapse ideology that has infused world culture. The willful ignorance of climate science around the planet has put us on the path to collapse, and any actions we take to alleviate collapse have to recognize how ubiquitous and insidious this ideology has become.

One group, called "Extinction Rebellion," has made the prospect of human extinction a part of its name.[233] Although its website has a more positive tone than the name indicates, it does employ the idea that human extinction from climate change is a distinct possibility.

At times pessimism about human survival becomes a bit flippant. One author, John Gray, renames *Homo Sapiens "Homo rapiens."* He writes, "Homo rapiens is only one of very many species, and not obviously worth preserving. Later or sooner, it will become extinct. When it is gone the Earth will recover. ... The earth will forget mankind."[234] *"Not worth preserving"* is a rather harsh judgment about humanity, but if humanity cannot discard collapse ideology, it will deserve this severe rebuke. Will it deserve extinction?

Because of the threat of human extinction, more and more scientists are looking at the prospects for humanity in a climate-changed world. George Marshall cites some data that should give us pause: "The Future of Humanity Institute conducted a poll of academic experts on global risks. They gave an estimate of 19 percent probability that the human species will go extinct before the end of this century. *The Stern Review: The Economics of Climate Change* factored a 9.5 percent risk of extinction within the next century into its calculations." Marshall quotes a famous comedian: "Extinction fits neatly into an altogether more flippant and fatalistic narrative that it is too late to do anything. As the late comedian George Carlin put it: "Save the planet! What!? Are these ... people kidding me!? The planet isn't going anywhere. We are! We are going away, so pack your [stuff], folks. We wouldn't leave much of a trace either. Just another failed mutation, just another closed end biological mistake. The planet will shake us off like a bad case of fleas. A surface nuisance."[235]

Although Carlin's statement is somewhat flippant, it is thought-provoking. Are humans just another "failed mutation," a "closed end biological mistake?" We can control our own fate, but it remains to be seen if we have the ability to reverse our fossil fuel addiction. Most writers are not optimistic.

Ultimately, issues of human extinction in the context of climate change are rather nihilistic. Do humans really want to add more carbon to the atmosphere when it leads to collapse? Probably not, although one does wonder about how collapse ideologues can reconcile their ideology without nihilism. One author, Clive Hamilton, has posed the issue as a question of human advancement: "Sooner or later, though, the nihilistic chickens will come home to roost. Are we really willing to give up on the human story and its almost unimaginable accomplishments, to regard it all as nothing and our existence as a mere assemblage of molecules that came together for a short time to resist entropy?"[236]

He answers his own question with a "no," but the real question is what kind of human advancement can be sustained. *Progress* animates much of human behavior that will lead to collapse, and assuming that progress will continue on its present course is a fool's errand. Either humanity will change the way progress is measured, to a much more sustainable path, or humanity will continue on its present course to extinction.

A more scientific version of human extinction places it in the context of evolution: "The Earth's history is a long story of numerous species birthing, evolving and eventually going extinct. There is no manifest destiny for our species. There is no divine promise that humanity may not in the future follow in the footsteps of the dinosaurs. Our lives are not transcendent to Nature nor the multitude of other natural forces, animals, plants, microbes and other life forms and molecules upon which our existence depends."[237]

While we are a species that could transcend these limitations, the key is "evolving," and in the case of humans that is more than a physical process. We need to evolve our social systems to overcome collapse ideology and deal effectively with climate change. Whether we can do that is an open question. "Our greatest challenge is not climate. Our most significant challenge is to become greater than who we have been that has allowed us to produce this outcome that now challenges our very existence. Impending climate collapse is not the problem—it is the compelling invitation to our own evolution that can no longer be ignored."[238] Indeed, we are at a crucial point in our evolution, one that requires that we overcome petty differences and cast off ideological constraints that prevent our evolution to sustainable living.

Anthropocene or Anthropocide?

"Anthropocene" is the name of the current age of the earth coined by Paul Crutzen, and "Anthropocentric" is the label for the kind of thinking that characterizes the current age. Kate Davies has described the problems that anthropocentrism poses for human survival: "Anthropocentrism places human beings at the center of concern and asserts that our species is separate from and superior to all others, including the earth itself."[239] This kind of human hubris deflects us from full consideration of the consequences of our actions. As another one of the many species on earth, we are subject to all of the forces of nature including those destructive forces we have ourselves unleashed.

Scientists have studied the threats to nature from anthropogenic climate change for decades, but only recently have gained enough data to actually observe what is happening. ""In the early 1990s we only had hints that we could drive the climate system over tipping points. We now know we might actually be witnessing the start of a mass extinction that could lead to our wiping out as much as half the species on Earth," said Jonathan Overpeck, environment dean at University of Michigan.[240]

As a self-centered species, humanity is ignoring the forces of nature that can threaten us and a major threat is posed by climate disruption. It is a form of denial to ignore nature. When we ignore the need for change, we encounter the natural constraints that will determine our future. Ozarko says: "Species come and go, but in mass extinction events, species vanish in droves." She describes the irony in this mode of thinking: "Ironically, most humans are unable to think of themselves at the level of a species. But that's precisely what we are: just another species. Despite our delusions of grandeur, we are not immune from our own self-destruction. Despite the overwhelming denial from those who project animus on the harbingers of this somber truth, the incontrovertible evidence for our imminent demise is becoming increasingly clear."[241]

One of the most difficult things about climate change is the element of time. We recognize that changes are coming, but they seem so remote and abstract that we do not think about when we will be facing the most severe consequences. We will begin to see more effects of climate change soon, however, and the buffer of time we have assumed will begin to compress. Ozarko observes, "The accelerating Earth changes catalyzed by our short-sighted ways, means that Gaea is now robbing us of the buffer of time that has conditioned into us such passive complacency. When the rug of the buffer of time is pulled from beneath the feet of humanity, most people become hostile and angry. Life without a future orientation in a civilization that knows no presence is profoundly intimidating and disorienting."[242] This hostility and anger will be directed to those who raise the alarm, depriving people of the cushion of time they assumed that they had.

Dunlap and Brulle, who have done a number of studies of public opinion about climate change, raise the possibilities of extinction: "The speed and extent of mounting anthropogenic global forcings (e.g., energy production and consumption, population growth, resource consumption, habitat destruction and fragmentation) could generate a global state shift and mass extinctions."[243] Mass extinctions of animals and plants in the Anthropocene are a major topic of research, as for example in Elizabeth Kolbert's *The Sixth Extinction.* She notes "having freed ourselves from the constraints of evolution, humans nevertheless remain dependent on the earth's biological and geochemical systems. By disrupting these systems—cutting down tropical rainforests, altering the composition of the atmosphere, acidifying the oceans—we're putting our own survival in danger."[244]

Richard Heinberg has dramatized the issue of human extinction with terms such as blackout. He says that "Humanity has survived many previous energy crises, from the Pleistocene megafaunal extinctions up to the oil shocks of the 1970s. ... Without a coherent effort to proactively reduce energy consumption further while developing renewable sources, the decline of energy from coal toward the middle of the century will deliver a coup de grace to industrial civilization, making the maintenance of electrical grids problematic to impossible." He calls this a "blackout" because we will use up fossil fuels without using the time to create sufficient amounts of renewable energy, the energy of the future.[245]

While this seems a bit overwrought, it does suggest one aspect of climate change that could lead to our extinction: the overuse of resources. The threat of running out of oil, sometimes called "peak oil," has been overcome by technology such as fracking, but in at least one interpretation of how we use resources, we are running out of clean air. It is difficult to grasp the concept of "dirty air" when CO_2, NO_x and methane, the three most significant greenhouse gases, are invisible. Humans react most strongly to immediate, visible threats and while we may worry about particulate pollution we are less concerned about what we cannot see or smell. As greenhouse gas concentrations increase (CO_2 has risen from 280 parts per million before 1750 to 410 today), we are using up the atmosphere's ability to maintain a balanced carbon cycle. Without this balance our health and survival are threatened.

Energy in the Future

What about the argument that we can, in fact, convert all energy to renewables and avoid the worst effects of climate change? That argument was made by Mark

Delucchi and Mark Jacobsen, who projected that construction of millions of wind and solar installations would cost about $100 trillion, but that could be recovered with sale of electricity at market rates.[246] Why don't we embrace this plan? Ian Angus offers an explanation: "That plan may be good for Earth's future, but the energy status quo is essential for the profit system today, and that will always take precedence. If an environmental plan would undermine the class and power relationships that define fossil capitalism, even if it would prevent climate catastrophe, then all the rational argument in the world won't produce the political will to implement it."[247] One does not have to be a radical Marxist to understand this feature of collapse ideology. Growth is essential to capitalism; without it, capitalism would fail.

Inequality is also built into capitalism and recent years have revealed how unequal incomes have become, within and between societies. What we have come to realize that these features of capitalism are also features of collapse ideology, beliefs that lead to the end of social organization. As they lead to collapse, they will also mean the end of capitalism. As Bill McKibben noted: "The behavior of the oil companies, which have pulled off perhaps the most consequential deception in mankind's history, is a prime example… In the case of global warming, the culprit is fossil fuel, the most lucrative commodity on earth, and so the companies responsible took a different tack." They could have invested in renewable energies with their enormous profits, but chose to stay with their existing business plans and attack the science. Referring to some investigatory journalism, he noted, "A document uncovered by the L.A. *Times* showed that, a month after [James] Hansen's testimony, in 1988, an unnamed Exxon "public affairs manager" issued an internal memo recommending that the company "emphasize the uncertainty" in the scientific data about climate change."[248] This has been a constant theme of collapse ideology: if we do not have absolutely certain results, we can ignore the science and carry on as usual.

Humans are particularly likely to overuse resources because they have been so successful in using them so far. Yuval Harari notes, "We humans have conquered the world thanks to our ability to create and believe fictional stories. We are therefore particularly bad at knowing the difference between fiction and reality."[249] We continue to believe the fictional story that we can overcome nature and be free of its constraints. That belief will eventually kill us.

Thinking the Unthinkable

We have learned that there are immediate existential threats to humanity, but we still resist thinking about solving them. As Spratt and Dunlap note in the report *What Lies Beneath*, "Often blind eyes were turned, either because of a lack of will to believe the signs, or an active preference to deny and then not to engage." They note that" "These deficiencies are clearly evident at the upper levels of climate policymaking, nationally and globally. They must be corrected as a matter of extreme urgency."[250]

It is ridiculous to argue that people can clearly see a threat and the ways to meet it, and then do nothing. However ridiculous it may seem, that is in fact what has happened in many countries. While most of the world leaders have at least agreed to some actions in the Paris Agreement, the societies they lead have not yet followed through.

As we think the unthinkable – about extinction, collective suicide and an uninhabitable earth – we may be led to a kind of nihilism. Indeed, nihilism has become a theme of modern history, from Al Qaeda to the Islamic State to predatory capitalism, where no morality constrains growth. Collapse is actually expected in these nihilistic movements. Although they may have short-term aspirations for more power or wealth, they are certainly well enough informed to realize that their long-term effect is destruction of society or the planet. Yet they proceed on their destructive courses without regard to the welfare of the rest of the world's population.

When we entered the Anthropocene around 1750 we started on the course of destabilizing the natural carbon cycle. The ability of the biosphere and the lithosphere to absorb carbon has been overwhelmed by the rapid buildup of greenhouse gases. As humans we have co-existed with nature for millennia, but as industrial creatures we are no longer collaborating with the natural cycles. Will we be able to restore our collaborative relationship with nature? Our survival as a species depends on our answer.

Why do the nations so furiously rage together? Why do the people imagine a vain thing?

Psalm II, Handel's Messiah No. 40

Appendix

IPCC 2018

Summary for Policymakers
http://report.ipcc.ch/sr15/pdf/sr15_spm_final.pdf

Introduction

This report responds to the invitation for IPCC '... to provide a Special Report in 2018 on the impacts of global warming of 1.5°C above pre-industrial levels and related global greenhouse gas emission pathways' contained in the Decision of the 21st Conference of Parties of the United Nations Framework Convention on Climate Change to adopt the Paris Agreement.[1]

The IPCC accepted the invitation in April 2016, deciding to prepare this Special Report on the impacts of global warming of 1.5°C above pre-industrial levels and related global greenhouse gas emission pathways, in the context of strengthening the global response to the threat of climate change, sustainable development, and efforts to eradicate poverty.

This Summary for Policy Makers (SPM) presents the key findings of the Special Report, based on the assessment of the available scientific, technical and socio-economic literature[2] relevant to global warming of 1.5°C and for the comparison between global warming of 1.5°C and 2°C above pre- industrial levels. The level of confidence associated with each key finding is reported using the IPCC calibrated language.[3] The underlying scientific basis of each key finding is indicated by references provided to chapter elements. In the SPM, knowledge gaps are identified associated with the underlying chapters of the report.

---

[1] Decision 1/CP.21, paragraph 21.
[2] The assessment covers literature accepted for publication by 15 May 2018.

[3] Each finding is grounded in an evaluation of underlying evidence and agreement. A level of confidence is expressed using five qualifiers: very low, low, medium, high and very high, and typeset in italics, for example, *medium confidence*. The following terms have been used to indicate the assessed likelihood of an outcome or a result: virtually certain 99–100% probability, very likely 90–100%, likely 66–100%, about as likely as not 33–66%, unlikely 0–33%, very unlikely 0–10%, exceptionally unlikely 0–1%. Additional terms (extremely likely 95–100%, more likely than not >50–100%, more unlikely than likely 0–<50%, extremely unlikely 0–5%) may also be used when appropriate. Assessed likelihood is typeset in italics, for example, *very likely*. This is consistent with AR5.

A. Understanding Global Warming of 1.5°C[4]

A1. Human activities are estimated to have caused approximately 1.0°C of global warming[5] above pre-industrial levels, with a *likely* range of 0.8°C to 1.2°C. Global warming is *likely* to reach 1.5°C between 2030 and 2052 if it continues to increase at the current rate. (*high confidence*) {1.2, Figure SPM.1}

A1.1. Reflecting the long-term warming trend since pre-industrial times, observed global mean surface temperature (GMST) for the decade 2006–2015 was 0.87°C (*likely* between 0.75°C and 0.99°C)[6] higher than the average over the 1850–1900 period (*very high confidence*). Estimated anthropogenic global warming matches the level of observed warming to within ±20% (*likely* range). Estimated anthropogenic global warming is currently increasing at 0.2°C (*likely* between 0.1°C and 0.3°C) per decade due to past and ongoing emissions (*high confidence*). {1.2.1, Table 1.1, 1.2.4}

---

[4] SPM BOX.1: Core Concepts
[5] Present level of global warming is defined as the average of a 30-year period centered on 2017 assuming the recent rate of warming continues.
[6] This range spans the four available peer-reviewed estimates of the observed GMST change and also accounts for additional uncertainty due to possible short-term natural variability. {1.2.1, Table 1.1}

A1.2. Warming greater than the global annual average is being experienced in many land regions and seasons, including two to three times higher in the Arctic. Warming is generally higher over land than over the ocean. (*high confidence*) {1.2.1, 1.2.2, Figure 1.1, Figure 1.3, 3.3.1, 3.3.2}

A1.3. Trends in intensity and frequency of some climate and weather extremes have been detected over time spans during which about 0.5°C of global warming occurred (*medium confidence*). This assessment is based on several lines of evidence, including attribution studies for changes in extremes since 1950. {3.3.1, 3.3.2, 3.3.3}

A.2. Warming from anthropogenic emissions from the pre-industrial period to the present will persist for centuries to millennia and will continue to cause further long-term changes in the climate system, such as sea level rise, with associated impacts (*high confidence*), but these emissions alone are *unlikely* to cause global warming of 1.5°C (*medium confidence*) {1.2, 3.3, Figure 1.5, Figure SPM.1}

A2.1. Anthropogenic emissions (including greenhouse gases, aerosols and their precursors) up to the present are *unlikely* to cause further warming of more than 0.5°C over the next two to three decades (*high confidence*) or on a century time scale (*medium confidence*). {1.2.4, Figure 1.5}

A2.2. Reaching and sustaining net-zero global anthropogenic CO_2 emissions and declining net non-CO_2 radiative forcing would halt anthropogenic global warming on multi-decadal timescales (*high confidence*). The maximum temperature reached is then determined by cumulative net global anthropogenic CO_2 emissions up to the time of net zero CO_2 emissions (*high confidence*) and the level of non-CO_2 radiative forcing in the decades prior to the time that maximum temperatures are reached (*medium confidence*). On longer timescales, sustained net negative global anthropogenic CO_2 emissions and/or further reductions in non-CO_2 radiative forcing may still be required to prevent further warming due to Earth system feedbacks and reverse ocean acidification (*medium confidence*) and will be required to minimise sea level rise (*high confidence*). {Cross-Chapter Box 2 in Chapter 1, 1.2.3, 1.2.4, Figure 1.4, 2.2.1, 2.2.2, 3.4.4.8, 3.4.5.1, 3.6.3.2}

Cumulative emissions of CO_2 and future non-CO_2 radiative forcing determine the probability of limiting warming to 1.5°C

a) Observed global temperature change and modeled responses to stylized anthropogenic emission and forcing pathways

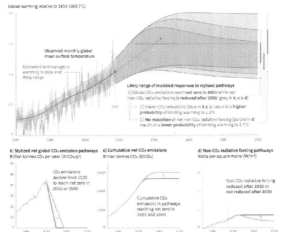

Observed monthly global mean surface temperature
Likely range of modeled responses to stylized pathways Global CO2 emissions reach **net zero in 2055** while net non-CO2 radiative forcing is **reduced a er 2030** (grey in **b**, **c** & **d**) Faster CO2 reductions (blue in **b** & **c**) result in a **higher probability** of limiting warming to 1.5°C

No reduction of net non-CO2 radiative forcing (purple in **d**) results in a **lower probability** of limiting warming to 1.5°C

Cumulative CO2 emissions in pathways reaching net zero in **2055** and **2040**
Faster immediate CO2 emission reductions limit cumulative CO2 emissions shown in panel **(c)**. Maximum temperature rise is determined by cumulative net CO2 emissions and net non-CO2 radiative forcing due to methane, nitrous oxide, aerosols and other anthropogenic forcing agents.

Figure SPM.1: Panel a: Observed monthly global mean surface temperature (GMST) change grey line up to 2017, from the HadCRUT4, GISTEMP, Cowtan–Way, and NOAA datasets) and estimated anthropogenic global warming (solid orange line up to 2017, with orange shading indicating assessed *likely* range). Orange dashed arrow and horizontal orange error bar show respectively central estimate and *likely* range of the time at which 1.5°C is reached if the current rate of warming continues. The grey plume on the right of Panel a) shows the *likely* range of warming responses, computed with a simple climate model, to a stylized pathway (hypothetical future) in which net CO_2 emissions (grey line in panels b and c) decline in a straight line from 2020 to reach net zero in 2055 and net non-CO_2 radiative forcing (grey line in panel d) increases to 2030 and then declines. The blue plume in panel a) shows the response to faster CO_2 emissions reductions (blue line in panel b), reaching net zero in 2040, reducing cumulative CO_2 emissions (panel c). The purple plume shows the response to net CO_2 emissions declining to zero in 2055, with net non-CO_2 forcing remaining constant after 2030. The vertical error bars on right of panel a) show the *likely* ranges (thin lines) and central terciles (33rd – 66th percentiles, thick lines) of the estimated distribution of warming in 2100 under these three stylized pathways. Vertical dotted error bars in panels b, c and d show the *likely* range of historical annual and cumulative global net CO_2 emissions in 2017 (data from the Global Carbon Project) and of net non-CO_2 radiative forcing in 2011 from AR5, respectively. Vertical axes in panels c and d are scaled to represent approximately equal effects on GMST. {1.2.1, 1.2.3, 1.2.4, 2.3, Chapter 1 Figure 1.2 & Chapter 1 Supplementary Material, Cross-Chapter Box 2}

A3. Climate-related risks for natural and human systems are higher for global warming of 1.5°C than at present, but lower than at 2°C (*high confidence*). These risks depend on the magnitude and rate of warming, geographic location, levels of development and vulnerability, and on the choices and implementation of adaptation and mitigation options (*high confidence*) (Figure SPM.2). {1.3, 3.3, 3.4, 5.6}

A3.1. Impacts on natural and human systems from global warming have already been observed (*high confidence*). Many land and ocean ecosystems and some of the services they provide have already changed due to global warming (*high confidence*). {1.4, 3.4, 3.5, Figure SPM.2}

A3.2. Future climate-related risks depend on the rate, peak and duration of warming. In the aggregate they are larger if global warming exceeds 1.5°C before returning to that level by 2100 than if global warming gradually stabilizes at 1.5°C, especially if the peak temperature is high (e.g., about 2°C) (*high confidence*). Some impacts may be long-lasting or irreversible, such as the loss of some ecosystems (*high confidence*). {3.2, 3.4.4, 3.6.3, Cross-Chapter Box 8}

A3.3. Adaptation and mitigation are already occurring (*high confidence*). Future climate-related risks would be reduced by the upscaling and acceleration of far-reaching, multi-level and cross- sectoral climate mitigation and by both incremental and transformational adaptation (*high confidence*). {1.2, 1.3, Table 3.5, 4.2.2, Cross-Chapter Box 9 in Chapter 4, Box 4.2, Box 4.3, Box 4.6, 4.3.1, 4.3.2, 4.3.3, 4.3.4, 4.3.5, 4.4.1, 4.4.4, 4.4.5, 4.5.3}

B. Projected Climate Change, Potential Impacts and Associated Risks

B1. Climate models project robust[7] differences in regional climate characteristics between present-day and global warming of 1.5°C,[8] and between 1.5°C and 2°C.[8] These differences include increases in: mean temperature in most

land and ocean regions (*high confidence*), hot extremes in most inhabited regions (*high confidence*), heavy precipitation in several regions (*medium confidence*), and the probability of drought and precipitation deficits in some regions (*medium confidence*). {3.3}

B1.1. Evidence from attributed changes in some climate and weather extremes for a global warming of about 0.5°C supports the assessment that an additional 0.5°C of warming compared to present is associated with further detectable changes in these extremes (*medium confidence*). Several regional changes in climate are assessed to occur with global warming up to 1.5°C compared to pre- industrial levels, including warming of extreme temperatures in many regions (*high confidence*), increases in frequency, intensity, and/or amount of heavy precipitation in several regions (*high confidence*), and an increase in intensity or frequency of droughts in some regions (*medium confidence*). {3.2, 3.3.1, 3.3.2, 3.3.3, 3.3.4, Table 3.2}

B1.2. Temperature extremes on land are projected to warm more than GMST (*high confidence*): extreme hot days in mid-latitudes warm by up to about 3°C at global warming of 1.5°C and about 4°C at 2°C, and extreme cold nights in high latitudes warm by up to about 4.5°C at 1.5°C and about 6°C at 2°C (*high confidence*). The number of hot days is projected to increase in most land regions, with highest increases in the tropics (*high confidence*). {3.3.1, 3.3.2, Cross-Chapter Box 8 in Chapter 3}

B1.3. Risks from droughts and precipitation deficits are projected to be higher at 2°C compared to 1.5°C global warming in some regions (*medium confidence*). Risks from heavy precipitation events are projected to be higher at 2°C compared to 1.5°C global warming in several northern hemisphere high-latitude and/or high-elevation regions, eastern Asia and eastern North America (*medium confidence*). Heavy precipitation associated with tropical cyclones is projected to be higher at 2°C compared to 1.5°C global warming (*medium confidence*). There is generally *low confidence* in projected changes in heavy precipitation at 2°C compared to 1.5°C in other regions. Heavy precipitation when aggregated at global scale is projected to be higher at 2.0°C than at 1.5°C of global warming (*medium confidence*). As a consequence of heavy precipitation, the fraction of the global land area affected by flood hazards is projected to be larger at 2°C compared to 1.5°C of global warming (*medium confidence*). {3.3.1, 3.3.3, 3.3.4, 3.3.5, 3.3.6}

B2. By 2100, global mean sea level rise is projected to be around 0.1 metre lower with global warming of 1.5°C compared to 2°C (*medium confidence*). Sea level will continue to rise well beyond 2100 (*high confidence*), and the magnitude and rate of this rise depends on future emission pathways. A slower rate of sea level rise enables greater opportunities for adaptation in the human and ecological systems of small islands, low-lying coastal areas and deltas (*medium confidence*). {3.3, 3.4, 3.6 }

B2.1. Model-based projections of global mean sea level rise (relative to 1986-2005) suggest an indicative range of 0.26 to 0.77 m by 2100 for 1.5°C global warming, 0.1 m (0.04-0.16 m) less than for a global warming of 2°C (*medium confidence*). A reduction of 0.1 m in global sea level rise implies that up to 10 million fewer people would be exposed to related risks, based on population in the year 2010 and assuming no adaptation (*medium confidence*). {3.4.4, 3.4.5, 4.3.2}

B2.2. Sea level rise will continue beyond 2100 even if global warming is limited to 1.5°C in the 21st century (*high confidence*). Marine ice sheet instability in Antarctica and/or irreversible loss of the Greenland ice sheet could result in multi-metre rise in sea level over hundreds to thousands of years.

These instabilities could be triggered around 1.5°C to 2°C of global warming (*medium confidence*). {3.3.9, 3.4.5, 3.5.2, 3.6.3, Box 3.3, Figure SPM.2}

B2.3. Increasing warming amplifies the exposure of small islands, low-lying coastal areas and deltas to the risks associated with sea level rise for many human and ecological systems, including increased saltwater intrusion, flooding and damage to infrastructure (*high confidence*). Risks associated with sea level rise are higher at 2°C compared to 1.5°C. The slower rate of sea level rise at global warming of 1.5°C reduces these risks enabling greater opportunities for adaptation including managing and restoring natural coastal ecosystems, and infrastructure reinforcement (*medium confidence*). {3.4.5, Figure SPM.2, Box 3.5}

B3. On land, impacts on biodiversity and ecosystems, including species loss and extinction, are projected to be lower at 1.5°C of global warming compared to 2°C. Limiting global warming to 1.5°C compared to 2°C is projected to lower the impacts on terrestrial, freshwater, and coastal ecosystems and to retain more of their services to humans (*high confidence*). (Figure SPM.2) {3.4, 3.5, Box

3.4, Box 4.2, Cross-Chapter Box 8 in Chapter 3}

B3.1. Of 105,000 species studied, 6% of insects, 8% of plants and 4% of vertebrates are projected to lose over half of their climatically determined geographic range for global warming of 1.5°C, compared with 18% of insects, 16% of plants and 8% of vertebrates for global warming of 2°C (*medium confidence*). Impacts associated with other biodiversity-related risks such as forest fires, and the spread of invasive species, are lower at 1.5°C compared to 2°C of global warming (*high confidence*). {3.4.3, 3.5.2}

B3.2. Approximately 4% (interquartile range 2–7%) of the global terrestrial land area is projected to undergo a transformation of ecosystems from one type to another at 1oC of global warming, compared with 13% (interquartile range 8–20%) at 2°C (*medium confidence*). This indicates that the area at risk is projected to be approximately 50% lower at 1.5°C compared to 2°C (*medium confidence*). {3.4.3.1, 3.4.3.5}

B3.3. High-latitude tundra and boreal forests are particularly at risk of climate change-induced degradation and loss, with woody shrubs already encroaching into the tundra (*high confidence*) and will proceed with further warming. Limiting global warming to 1.5°C rather than 2°C is projected to prevent the thawing over centuries of a permafrost area in the range of 1.5 to 2.5 million km^2 (*medium confidence*). {3.3.2, 3.4.3, 3.5.5}

B4. Limiting global warming to 1.5°C compared to 2°C is projected to reduce increases in ocean temperature as well as associated increases in ocean acidity and decreases in ocean oxygen levels (*high confidence*). Consequently, limiting global warming to 1.5°C is projected to reduce risks to marine biodiversity, fisheries, and ecosystems, and their functions and services to humans, as illustrated by recent changes to Arctic sea ice and warm water coral reef ecosystems (*high confidence*). {3.3, 3.4, 3.5, Boxes 3.4, 3.5}

B4.1. There is *high confidence* that the probability of a sea-ice-free Arctic Ocean during summer is substantially lower at global warming of 1.5°C when compared to 2°C. With 1.5°C of global warming, one sea ice-free Arctic summer is projected per century. This likelihood is increased to at least one per decade with 2°C global warming. Effects of a temperature overshoot are reversible for Arctic sea ice cover on decadal time scales (*high confidence*). {3.3.8, 3.4.4.7}

B4.2. Global warming of 1.5°C is projected to shift the ranges of many marine species, to higher latitudes as well as increase the amount of damage to many ecosystems. It is also expected to drive the loss of coastal resources, and reduce the productivity of fisheries and aquaculture (especially at low latitudes). The risks of climate-induced impacts are projected to be higher at 2°C than those at global warming of 1.5°C (*high confidence*). Coral reefs, for example, are projected to decline by a further 70–90% at 1.5°C (*high confidence*) with larger losses (>99%) at 2oC (*very high confidence*). The risk of irreversible loss of many marine and coastal ecosystems increases with global warming, especially at 2°C or more (*high confidence*). {3.4.4, Box 3.4}

B4.3. The level of ocean acidification due to increasing CO_2 concentrations associated with global warming of 1.5°C is projected to amplify the adverse effects of warming, and even further at 2°C, impacting the growth, development, calcification, survival, and thus abundance of a broad range of species, e.g., from algae to fish (*high confidence*). {3.3.10, 3.4.4}

B4.4. Impacts of climate change in the ocean are increasing risks to fisheries and aquaculture via impacts on the physiology, survivorship, habitat, reproduction, disease incidence, and risk of invasive species (*medium confidence*) but are projected to be less at 1.5oC of global warming than at 2oC. One global fishery model, for example, projected a decrease in global annual catch for marine fisheries of about 1.5 million tonnes for 1.5°C of global warming compared to a loss of more than 3 million tonnes for 2°C of global warming (*medium confidence*). {3.4.4, Box 3.4}

B5. Climate-related risks to health, livelihoods, food security, water supply, human security, and economic growth are projected to increase with global warming of 1.5°C and increase further with 2°C. (Figure SPM.2) {3.4, 3.5, 5.2, Box 3.2, Box 3.3, Box 3.5, Box 3.6, Cross- Chapter Box 6 in Chapter 3, Cross-Chapter Box 9 in Chapter 4, Cross-Chapter Box 12 in Chapter 5, 5.2}

B5.1. Populations at disproportionately higher risk of adverse consequences of global warming of 1.5°C and beyond include disadvantaged and vulnerable populations, some indigenous peoples, and local communities dependent on agricultural or coastal livelihoods (*high confidence*). Regions at disproportionately higher risk include Arctic ecosystems, dryland regions, small-island developing states, and least developed countries (*high confidence*).

Poverty and disadvantages are expected to increase in some populations as global warming increases; limiting global warming to 1.5°C, compared with 2°C, could reduce the number of people both exposed to climate-related risks and susceptible to poverty by up to several hundred million by 2050 (*medium confidence*). {3.4.10, 3.4.11, Box 3.5, Cross-Chapter Box 6 in Chapter 3, Cross-Chapter Box 9 in Chapter 4, Cross-Chapter Box 12 in Chapter 5, 4.2.2.2, 5.2.1, 5.2.2, 5.2.3, 5.6.3}

B5.2. Any increase in global warming is projected to affect human health, with primarily negative consequences (*high confidence*). Lower risks are projected at 1.5°C than at 2°C for heat-related morbidity and mortality (*very high confidence*) and for ozone-related mortality if emissions needed for ozone formation remain high (*high confidence*). Urban heat islands often amplify the impacts of heatwaves in cities (*high confidence*). Risks from some vector-borne diseases, such as malaria and dengue fever, are projected to increase with warming from 1.5°C to 2°C, including potential shifts in their geographic range (*high confidence*). {3.4.7, 3.4.8, 3.5.5.8}

B5.3. Limiting warming to 1.5°C, compared with 2°C, is projected to result in smaller net reductions in yields of maize, rice, wheat, and potentially other cereal crops, particularly in sub- Saharan Africa, Southeast Asia, and Central and South America; and in the CO_2 dependent, nutritional quality of rice and wheat (*high confidence*). Reductions in projected food availability are larger at 2°C than at 1.5°C of global warming in the Sahel, southern Africa, the Mediterranean, central Europe, and the Amazon (*medium confidence*). Livestock are projected to be adversely affected with rising temperatures, depending on the extent of changes in feed quality, spread of diseases, and water resource availability (*high confidence*). {3.4.6, 3.5.4, 3.5.5, Box 3.1, Cross- Chapter Box 6 in Chapter 3, Cross-Chapter Box 9 in Chapter 4}

B5.4. Depending on future socioeconomic conditions, limiting global warming to 1.5°C, compared to 2°C, may reduce the proportion of the world population exposed to a climate-change induced increase in water stress by up to 50%, although there is considerable variability between regions (*medium confidence*). Many small island developing states would experience lower water stress as a result of projected changes in aridity when global warming is limited to 1.5°C, as compared to 2°C (*medium confidence*). {3.3.5, 3.4.2, 3.4.8, 3.5.5, Box

3.2, Box 3.5, Cross-Chapter Box 9 in Chapter 4}

B5.5. Risks to global aggregated economic growth due to climate change impacts are projected to be lower at 1.5°C than at 2°C by the end of this century[10] (*medium confidence*). This excludes the costs of mitigation, adaptation investments and the benefits of adaptation. Countries in the tropics and Southern Hemisphere subtropics are projected to experience the largest impacts on economic growth due to climate change should global warming increase from 1.5°C to 2 °C (*medium confidence*). {3.5.2, 3.5.3}

B5.6. Exposure to multiple and compound climate-related risks increases between 1.5°C and 2°C of global warming, with greater proportions of people both so exposed and susceptible to poverty in Africa and Asia (*high confidence*). For global warming from 1.5°C to 2°C, risks across energy, food, and water sectors could overlap spatially and temporally, creating new and exacerbating current hazards, exposures, and vulnerabilities that could affect increasing numbers of people and regions (*medium confidence*). {Box 3.5, 3.3.1, 3.4.5.3, 3.4.5.6, 3.4.11, 3.5.4.9}

B5.7. There are multiple lines of evidence that since the AR5 the assessed levels of risk increased for four of the five Reasons for Concern (RFCs) for global warming to 2°C (*high confidence*). The risk transitions by degrees of global warming are now: from high to very high between 1.5°C and 2°C for RFC1 (Unique and threatened systems) (*high confidence*); from moderate to high risk between 1.0°C and 1.5°C for RFC2 (Extreme weather events) (*medium confidence*); from moderate to high risk between 1.5°C and 2°C for RFC3 (Distribution of impacts) (*high confidence*); from moderate to high risk between 1.5°C and 2.5°C for RFC4 (Global aggregate impacts) (*medium confidence*); and from moderate to high risk between 1°C and 2.5°C for RFC5 (Large-scale singular events) (*medium confidence*). (Figure SPM.2) {3.4.13; 3.5, 3.5.2}

How the level of global warming affects impacts and/or risks associated with the Reasons for Concern (RFCs) and selected natural, managed and human systems

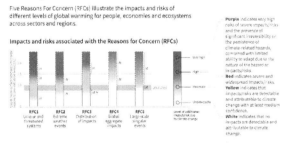

[10] Here, impacts on economic growth refer to changes in GDP. Many impacts, such as loss of human lives, cultural heritage, and ecosystem services, are difficult to value and monetize.

Five Reasons For Concern (RFCs) illustrate the impacts and risks of different levels of global warming for people, economies and ecosystems across sectors and regions.

Purple indicates very high risks of severe impacts/risks and the presence of significant irreversibility or the persistence of climate-related hazards, combined with limited ability to adapt due to the nature of the hazard or impacts/risks.

Red indicates severe and widespread impacts/risks.
Yellow indicates that impacts/risks are detectable and attributable to climate change with at least medium confidence.

White indicates that no impacts are detectable and attributable to climate change

Impacts and risks associated with the Reasons for Concern (RFCs)

Impacts and risks for selected natural, managed and human systems

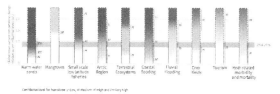

Confidence level for transition: *L*=Low, *M*=Medium, *H*=High and *VH*=Very high

Global mean surface temperature change Global mean surface temperature change relative to pre-industrial levels (oC) relative to pre-industrial levels (oC)

Figure SPM.2: Five integrative reasons for concern (RFCs) provide a framework for summarizing key impacts and risks across sectors and regions, and were introduced in the IPCC Third Assessment Report. RFCs illustrate the implications of global warming for people, economies, and ecosystems. Impacts and/or risks for each RFC are based on assessment of the new literature that has appeared. As in the AR5, this literature was used to make expert judgments to assess the levels of global warming at which levels of impact and/or risk are undetectable, moderate, high or very high. The selection of impacts and risks to natural, managed and human systems in the lower panel is illustrative and is not intended to be fully comprehensive. **RFC1 Unique and threatened systems:** ecological and human systems that have restricted geographic ranges constrained by climate related conditions and have high endemism or other distinctive properties. Examples include coral reefs, the Arctic and its indigenous people, mountain glaciers, and biodiversity hotspots. **RFC2 Extreme weather events**: risks/impacts to human health, livelihoods, assets, and ecosystems from extreme weather events such as heat waves, heavy rain, drought and associated wildfires, and coastal flooding. **RFC3 Distribution of impacts:** risks/impacts that disproportionately affect particular groups due to uneven distribution of physical climate change hazards, exposure or vulnerability. **RFC4 Global aggregate impacts:** global monetary damage, global scale degradation and loss of ecosystems and biodiversity. **RFC5 Large-scale singular events:** are relatively large, abrupt and sometimes irreversible changes in systems that are caused by global warming. Examples include disintegration of the Greenland and Antarctic ice sheets. {3.4, 3.5, 3.5.2.1, 3.5.2.2, 3.5.2.3, 3.5.2.4, 3.5.2.5, 5.4.1 5.5.3, 5.6.1, Box 3.4}

B6. There are limits to adaptation and adaptive capacity for some human and natural systems at global warming of 1.5°C, with associated losses (*medium confidence*). The number and availability of adaptation options vary by sector (*medium confidence*). {Table 3.5, 4.3, 4.5,

Cross-Chapter Box 9 in Chapter 4, Cross-Chapter Box 12 in Chapter 5}

B6.1. A wide range of adaptation options are available to reduce the risks to natural and managed ecosystems (e.g., ecosystem-based adaptation, ecosystem restoration and avoided degradation and deforestation, biodiversity management, sustainable aquaculture, and local knowledge and indigenous knowledge), the risks of sea level rise (e.g., coastal defence and hardening), and the risks to health, livelihoods, food, water, and economic growth, especially in rural landscapes (e.g., efficient irrigation, social safety nets, disaster risk management, risk spreading and sharing, community-based adaptation) and urban areas (e.g., green infrastructure, sustainable land use and planning, and sustainable water management) (*medium confidence*). {4.3.1, 4.3.2, 4.3.3, 4.3.5, 4.5.3, 4.5.4, 5.3.2, Box 4.2, Box 4.3, Box 4.6, Cross-Chapter Box 9 in Chapter 4}.

B6.2. Adaptation is expected to be more challenging for ecosystems, food and health systems at 2°C of global warming than for 1.5°C (*medium confidence*). Some vulnerable regions, including small islands and Least Developed Countries, are projected to experience high multiple interrelated climate risks even at global warming of 1.5°C (*high confidence*). {3.3.1, 3.4.5, Box 3.5, Table 3.5, Cross-Chapter Box 9 in Chapter 4, 5.6, Cross-Chapter Box 12 in Chapter 5, Box 5.3}

B6.3. Limits to adaptive capacity exist at 1.5°C of global warming, become more pronounced at higher levels of warming and vary by sector, with site-specific implications for vulnerable regions, ecosystems, and human health (*medium confidence*) {Cross-Chapter Box 12 in Chapter 5, Box 3.5, Table 3.5}

C. Emission Pathways and System Transitions Consistent with 1.5°C Global Warming

C1. In model pathways with no or limited overshoot of 1.5°C, global net anthropogenic CO_2 emissions decline by about 45% from 2010 levels by 2030 (40–60% interquartile range), reaching net zero around 2050 (2045–2055 interquartile range). For limiting global warming to below 2°C[11] CO_2 emissions are projected to decline by about 20% by 2030 in most pathways (10–30% interquartile range) and reach net zero around 2075 (2065–2080 interquartile range). Non-CO_2 emissions in pathways that limit global warming to 1.5°C show deep reductions that are similar to those

in pathways limiting warming to 2°C. (*high confidence*) (Figure SPM.3a) {2.1, 2.3, Table 2.4}

C1.1. CO_2 emissions reductions that limit global warming to 1.5°C with no or limited overshoot can involve different portfolios of mitigation measures, striking different balances between lowering energy and resource intensity, rate of decarbonization, and the reliance on carbon dioxide removal. Different portfolios face different implementation challenges, and potential synergies and trade-offs with sustainable development. (*high confidence*). (Figure SPM.3b) {2.3.2, 2.3.4, 2.4, 2.5.3}

[11] References to pathways limiting global warming to 2°C are based on a 66% probability of staying below 2°C.

Most adaptation needs will be lower for global warming of 1.5°C compared to 2°C (*high confidence*). There are a wide range of adaptation options that can reduce the risks of climate change (*high confidence*).

C1.2. Modelled pathways that limit global warming to 1.5°C with no or limited overshoot involve deep reductions in emissions of methane and black carbon (35% or more of both by 2050 relative to 2010). These pathways also reduce most of the cooling aerosols, which partially offsets mitigation effects for two to three decades. Non-CO_2 emissions[12] can be reduced as a result of broad mitigation measures in the energy sector. In addition, targeted non-CO_2 mitigation measures can reduce nitrous oxide and methane from agriculture, methane from the waste sector, some sources of black carbon, and hydrofluorocarbons. High bioenergy demand can increase emissions of nitrous oxide in some 1.5°C pathways, highlighting the importance of appropriate management approaches. Improved air quality resulting from projected reductions in many non-CO_2 emissions provide direct and immediate population health benefits in all 1.5°C model pathways. (*high confidence*) (Figure SPM.3a) {2.2.1, 2.3.3, 2.4.4, 2.5.3, 4.3.6, 5.4.2}

C1.3. Limiting global warming requires limiting the total cumulative global anthropogenic emissions of CO_2 since the preindustrial period, i.e. staying within a total carbon budget (*high confidence*).[13] By the end of 2017, anthropogenic CO_2 emissions since the preindustrial period are estimated to have reduced the total carbon budget for 1.5°C by approximately 2200 ± 320 $GtCO_2$ (*medium confidence*). The associated remaining budget is being depleted by current emissions of 42 ± 3

$GtCO_2$ per year (*high confidence*). The choice of the measure of global temperature affects the estimated remaining carbon budget. Using global mean surface air temperature, as in AR5, gives an estimate of the remaining carbon budget of 580 $GtCO_2$ for a 50% probability of limiting warming to 1.5°C, and 420 $GtCO_2$ for a 66% probability (*medium confidence*).[14] Alternatively, using GMST gives estimates of 770 and 570 $GtCO_2$, for 50% and 66% probabilities,[15] respectively (*medium confidence*). Uncertainties in the size of these estimated remaining carbon budgets are substantial and depend on several factors. Uncertainties in the climate response to CO_2 and non-CO_2 emissions contribute ±400 $GtCO_2$ and the level of historic warming contributes ±250 $GtCO_2$ (*medium confidence*). Potential additional carbon release from future permafrost thawing and methane release from wetlands would reduce budgets by up to 100 $GtCO_2$ over the course of this century and more thereafter (*medium confidence*). In addition, the level of non-CO_2 mitigation in the future could alter the remaining carbon budget by 250 $GtCO_2$ in either direction (*medium confidence*). {1.2.4, 2.2.2, 2.6.1, Table 2.2, Chapter 2 Supplementary Material}

C1.4. Solar radiation modification (SRM) measures are not included in any of the available assessed pathways. Although some SRM measures may be theoretically effective in reducing an overshoot, they face large uncertainties and knowledge gaps as well as substantial risks,

[12] Non-CO_2 emissions included in this report are all anthropogenic emissions other than CO_2 that result in radiative forcing. These include short-lived climate forcers, such as methane, some fluorinated gases, ozone precursors, aerosols or aerosol precursors, such as black carbon and sulphur dioxide, respectively, as well as long-lived greenhouse gases, such as nitrous oxide or some fluorinated gases. The radiative forcing associated with non-CO_2 emissions and changes in surface albedo is referred to as non-CO_2 radiative forcing. {x.y}

[13] There is a clear scientific basis for a total carbon budget consistent with limiting global warming to 1.5°C. However, neither this total carbon budget nor the fraction of this budget taken up by past emissions were assessed in this report.

[14] Irrespective of the measure of global temperature used, updated understanding and further advances in methods have led to an increase in the estimated remaining carbon budget of about 300 $GtCO_2$ compared to AR5. (*medium confidence*) {x.y}

[15] These estimates use observed GMST to 2006–2015 and estimate future temperature changes using near surface air temperatures.

institutional and social constraints to deployment related to governance, ethics, and impacts on sustainable development. They also do not mitigate ocean acidification. (*medium confidence*). {4.3.8, Cross-Chapter Box 10 in Chapter 4}

Global emissions pathway characteristics

General characteristics of the evolution of anthropogenic net emissions of CO2, and total emissions of methane, black carbon, and nitrous oxide in model pathways that limit global warming to 1.5°C with no or limited overshoot. Net emissions are defined as anthropogenic emissions reduced by anthropogenic removals. Reductions in net emissions can be achieved through different portfolios of mitigation measures illustrated in Figure SPM3B.

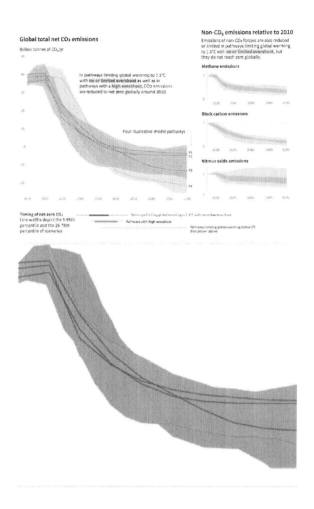

Figure SPM.3a: Global emissions pathway characteristics. The main panel shows global net anthropogenic CO_2 emissions in pathways limiting global warming to 1.5°C with no or limited (less than 0.1°C) overshoot and pathways with higher overshoot. The shaded area shows the full range for pathways analysed in this report. The panels on the right show non-CO_2 emissions ranges for three compounds with large historical forcing and a substantial portion of emissions coming from sources distinct from those central to CO_2 mitigation. Shaded areas in these panels show the 5–95% (light shading) and interquartile (dark shading) ranges of pathways limiting global warming to 1.5°C with no or limited overshoot. Box and whiskers at the bottom of the figure show the timing of pathways reaching global net zero CO_2 emission levels, and a comparison with pathways limiting global warming to 2°C with at least 66% probability. Four illustrative model pathways are highlighted in the main panel and are labelled P1, P2, P3 and P4, corresponding to the LED, S1, S2, and S5 pathways assessed in Chapter 2. Descriptions and characteristics of these pathways are available in Figure SPM3b. {2.1, 2.2, 2.3, Figure 2.5, Figure 2.10, Figure 2.11}

Characteristics of four illustrative model pathways

Different mitigation strategies can achieve the net emissions reductions that would be required to follow a pathway that limit global warming to 1.5°C with no or limited overshoot. All pathways use Carbon Dioxide Removal (CDR), but the amount varies across pathways, as do the relative contributions of Bioenergy with Carbon Capture and Storage (BECCS) and removals in the Agriculture, Forestry and Other Land Use (AFOLU) sector. This has implications for the emissions and several other pathway characteristics.

Breakdown of contributions to global net CO2 emissions in four illustrative model pathways

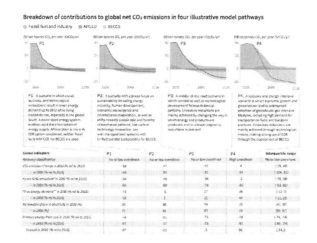

Figure SPM.3b: Characteristics of four illustrative model pathways in relation to global warming of 1.5°C introduced in Figure SPM3a. These pathways were selected to show a range of potential mitigation approaches and vary widely in their projected energy and land use, as well as their assumptions about future socioeconomic developments, including economic and population growth, equity and sustainability. A breakdown of the global net anthropogenic CO_2 emissions into the contributions in terms of CO_2 emissions from fossil fuel and industry, agriculture, forestry and other land use (AFOLU), and bioenergy with carbon capture and storage (BECCS) is shown. AFOLU estimates reported here are not necessarily comparable with countries' estimates. Further characteristics for each of these pathways are listed below each pathway. These pathways illustrate relative global differences in mitigation strategies, but do not represent central estimates, national strategies, and do not indicate requirements. For comparison, the right-most column shows the interquartile ranges across pathways with no or limited overshoot of 1.5°C. Pathways P1, P2, P3 and P4, correspond to the LED, S1, S2, and S5 pathways assessed in Chapter 2. (Figure SPM.3a) {2.2.1, 2.3.1, 2.3.2, 2.3.3, 2.3.4, 2.4.1, 2.4.2, 2.4.4, 2.5.3, Figure 2.5, Figure 2.6, Figure 2.9, Figure 2.10, Figure 2.11, Figure 2.14, Figure 2.15, Figure 2.16, Figure 2.17, Figure 2.24, Figure 2.25, Table 2.4, Table 2.6, Table 2.7, Table 2.9, Table 4.1}

C2. Pathways limiting global warming to 1.5°C with no or limited overshoot would require rapid and far-reaching transitions in energy, land, urban and infrastructure (including transport and buildings), and industrial systems (*high confidence*). These systems transitions are unprecedented in terms of scale, but not necessarily in terms of speed, and imply deep emissions reductions in all sectors, a wide portfolio of mitigation options and a significant upscaling of investments in those options (*medium confidence*). {2.3, 2.4, 2.5, 4.2, 4.3, 4.4, 4.5}

C2.1. Pathways that limit global warming to 1.5°C with no or limited overshoot show system changes that are more rapid and pronounced over the next two decades than in 2°C pathways (*high confidence*). The rates of system changes associated with limiting global warming to 1.5°C with no or limited overshoot have occurred in the past within specific sectors, technologies and spatial contexts, but there is no documented historic precedent for their scale (*medium confidence*). {2.3.3, 2.3.4, 2.4, 2.5, 4.2.1, 4.2.2, Cross-Chapter Box 11 in Chapter 4}

C2.2. In energy systems, modelled global pathways (considered in the literature) limiting global warming to 1.5°C with no or limited overshoot (for more details see

Figure SPM.3b), generally meet energy service demand with lower energy use, including through enhanced energy efficiency, and show faster electrification of energy end use compared to 2°C (*high confidence*). In 1.5°C pathways with no or limited overshoot, low-emission energy sources are projected to have a higher share, compared with 2°C pathways, particularly before 2050 (*high confidence*). In 1.5°C pathways with no or limited overshoot, renewables are projected to supply 70–85% (interquartile range) of electricity in 2050 (*high confidence*). In electricity generation, shares of nuclear and fossil fuels with carbon dioxide capture and storage (CCS) are modelled to increase in most 1.5°C pathways with no or limited overshoot. In modelled 1.5°C pathways with limited or no overshoot, the use of CCS would allow the electricity generation share of gas to be approximately 8% (3–11% interquartile range) of global electricity in 2050, while the use of coal shows a steep reduction in all pathways and would be reduced to close to 0% (0–2%) of electricity (*high confidence*). While acknowledging the challenges, and differences between the options and national circumstances, political, economic, social and technical feasibility of solar energy, wind energy and electricity storage technologies have substantially improved over the past few years (*high confidence*).

These improvements signal a potential system transition in electricity generation (Figure SPM.3b) {2.4.1, 2.4.2, Figure 2.1, Table 2.6, Table 2.7, Cross-Chapter Box 6 in Chapter 3, 4.2.1, 4.3.1, 4.3.3, 4.5.2}

C2.3. CO_2 emissions from industry in pathways limiting global warming to 1.5°C with no or limited overshoot are projected to be about 75–90% (interquartile range) lower in 2050 relative to 2010, as compared to 50–80% for global warming of 2°C (*medium confidence*). Such reductions can be achieved through combinations of new and existing technologies and practices, including electrification, hydrogen, sustainable bio-based feedstocks, product substitution, and carbon capture, utilization and storage (CCUS). These options are technically proven at various scales but their large-scale deployment may be limited by economic, financial, human capacity and institutional constraints in specific contexts, and specific characteristics of large-scale industrial installations. In industry, emissions reductions by energy and process efficiency by themselves are insufficient for limiting warming to 1.5°C with no or limited overshoot (*high confidence*). {2.4.3, 4.2.1, Table 4.1, Table 4.3, 4.3.3, 4.3.4, 4.5.2}

C2.4. The urban and infrastructure system transition consistent with limiting global warming to 1.5°C with no or limited overshoot would imply, for example, changes in land and urban planning practices, as well as deeper emissions reductions in transport and buildings compared to pathways that limit global warming below 2°C (see 2.4.3; 4.3.3; 4.2.1) (*medium confidence*). Technical measures and practices enabling deep emissions reductions include various energy efficiency options. In pathways limiting global warming to 1.5°C with no or limited overshoot, the electricity share of energy demand in buildings would be about 55–75% in 2050 compared to 50–70% in 2050 for 2°C global warming (*medium confidence*). In the transport sector, the share of low-emission final energy would rise from less than 5% in 2020 to about 35–65% in 2050 compared to 25–45% for 2°C global warming (*medium confidence*). Economic, institutional and socio-cultural barriers may inhibit these urban and infrastructure system transitions, depending on national, regional and local circumstances, capabilities and the availability of capital (*high confidence*). {2.3.4, 2.4.3, 4.2.1, Table 4.1, 4.3.3, 4.5.2}.

C2.5. Transitions in global and regional land use are found in all pathways limiting global warming to 1.5°C with no or limited overshoot, but their scale depends on the pursued mitigation portfolio. Model pathways that limit global warming to 1.5°C with no or limited overshoot project the conversion of 0.5–8 million km^2 of pasture and 0–5 million km^2 of non-pasture agricultural land for food and feed crops into 1–7 million km^2 for energy crops and a 1 million km^2 reduction to 10 million km^2 increase in forests by 2050 relative to 2010 (*medium confidence*).[16] Land use transitions of similar magnitude can be observed in modelled 2°C pathways (*medium confidence*). Such large transitions pose profound challenges for sustainable management of the various demands on land for human settlements, food, livestock feed, fibre, bioenergy, carbon storage, biodiversity and other ecosystem services (*high confidence*). Mitigation options limiting the demand for land include sustainable intensification of land use practices, ecosystem restoration and changes towards less resource-intensive diets (*high confidence*). The implementation of land-based mitigation options would require overcoming socio-economic, institutional, technological, financing and environmental barriers that differ across regions (*high confidence*). {2.4.4, Figure 2.24, 4.3.2, 4.5.2, Cross-Chapter Box 7 in Chapter 3}

C2.6 Total annual average energy-related mitigation investment for the period 2015 to 2050 in pathways limiting warming to 1.5°C is estimated to be around 900 billion USD2015 (range of 180 billion to 1800 billion USD2015 across six models[17]). This corresponds to total annual average energy supply investments of 1600 to 3800 billion USD2015 and total annual average energy demand investments of 700 to 1000 billion USD2015 for the period 2015 to 2050, and an increase in total energy-related investments of about 12% (range of 3% to 23%) in 1.5°C pathways relative to 2°C pathways. Average annual investment in low-carbon energy technologies and energy efficiency are upscaled by roughly a factor of five (range of factor of 4 to 5) by 2050 compared to 2015 (*medium confidence*). {2.5.2, Box 4.8, Figure 2.27}

C2.7. Modelled pathways limiting global warming to 1.5°C with no or limited overshoot project a wide range of global average discounted marginal abatement costs over the 21st century. They are roughly 3-4 times higher than in pathways limiting global warming to below 2°C (*high confidence*). The economic literature distinguishes marginal abatement costs from total mitigation costs in the economy. The literature on total mitigation costs of 1.5°C mitigation pathways is limited and was not assessed in this report. Knowledge gaps remain in the integrated assessment of the economy wide costs and benefits of mitigation in line with pathways limiting warming to 1.5°C. {2.5.2; 2.6; Figure 2.26}

[16] The projected land use changes presented are not deployed to their upper limits simultaneously in a single pathway.
[17] Including two pathways limiting warming to 1.5°C with no or limited overshoot and four pathways with high overshoot.

C3. All pathways that limit global warming to 1.5°C with limited or no overshoot project the use of carbon dioxide removal (CDR) on the order of 100–1000 GtCO$_2$ over the 21st century. CDR would be used to compensate for residual emissions and, in most cases, achieve net negative emissions to return global warming to 1.5°C following a peak (*high confidence*). CDR deployment of several hundreds of GtCO$_2$ is subject to multiple feasibility and sustainability constraints (*high confidence*). Significant near-term emissions reductions and measures to lower energy and land demand can limit CDR deployment to a few hundred GtCO$_2$ without reliance on bioenergy with carbon capture and storage (BECCS) (*high confidence*). {2.3, 2.4, 3.6.2, 4.3, 5.4}

C3.1. Existing and potential CDR measures include afforestation and reforestation, land restoration and soil carbon sequestration, BECCS, direct air carbon capture and storage (DACCS), enhanced weathering and ocean alkalinization. These differ widely in terms of maturity, potentials, costs, risks, co-benefits and trade-offs (*high confidence*). To date, only a few published pathways include CDR measures other than afforestation and BECCS. {2.3.4, 3.6.2, 4.3.2, 4.3.7}

C3.2. In pathways limiting global warming to 1.5°C with limited or no overshoot, BECCS deployment is projected to range from 0–1, 0–8, and 0–16 $GtCO_2 \, yr^{-1}$ in 2030, 2050, and 2100, respectively, while agriculture, forestry and land-use (AFOLU) related CDR measures are projected to remove 0–5, 1–11, and 1–5 $GtCO_2 \, yr^{-1}$ in these years (*medium confidence*). The upper end of these deployment ranges by mid-century exceeds the BECCS potential of up to 5 $GtCO_2 \, yr^{-1}$ and afforestation potential of up to 3.6 $GtCO_2 \, yr^{-1}$ assessed based on recent literature (*medium confidence*). Some pathways avoid BECCS deployment completely through demand-side measures and greater reliance on AFOLU-related CDR measures (*medium confidence*). The use of bioenergy can be as high or even higher when BECCS is excluded compared to when it is included due to its potential for replacing fossil fuels across sectors (*high confidence*). (Figure SPM.3b) {2.3.3, 2.3.4, 2.4.2, 3.6.2, 4.3.1, 4.2.3, 4.3.2, 4.3.7, 4.4.3, Table 2.4}

C3.3. Pathways that overshoot 1.5°C of global warming rely on CDR exceeding residual CO_2 emissions later in the century to return to below 1.5°C by 2100, with larger overshoots requiring greater amounts of CDR (Figure SPM.3b). (*high confidence*). Limitations on the speed, scale, and societal acceptability of CDR deployment hence determine the ability to return global warming to below 1.5°C following an overshoot. Carbon cycle and climate system understanding is still limited about the effectiveness of net negative emissions to reduce temperatures after they peak (*high confidence*). {2.2, 2.3.4, 2.3.5, 2.6, 4.3.7, 4.5.2, Table 4.11}

C3.4. Most current and potential CDR measures could have significant impacts on land, energy, water, or nutrients if deployed at large scale (*high confidence*). Afforestation and bioenergy may compete with other land uses and may have significant impacts on agricultural and food systems, biodiversity and other ecosystem functions and services (*high confidence*). Effective governance is needed to limit such trade-offs and ensure permanence of carbon removal in terrestrial, geological and ocean reservoirs (*high confidence*). Feasibility and sustainability of CDR use could be enhanced by a portfolio of options deployed at substantial, but lesser scales, rather than a single option at very large scale (*high confidence*). (Figure SPM.3b). {2.3.4, 2.4.4, 2.5.3, 2.6, 3.6.2, 4.3.2, 4.3.7, 4.5.2, 5.4.1, 5.4.2; Cross-Chapter Boxes 7 and 8 in Chapter 3, Table 4.11, Table 5.3, Figure 5.3}

C3.5. Some AFOLU-related CDR measures such as restoration of natural ecosystems and soil carbon sequestration could provide co-benefits such as improved biodiversity, soil quality, and local food security. If deployed at large scale, they would require governance systems enabling sustainable land management to conserve and protect land carbon stocks and other ecosystem functions and services (*medium confidence*). (Figure SPM.4) {2.3.3, 2.3.4, 2.4.2, 2.4.4, 3.6.2, 5.4.1, Cross-Chapter Boxes 3 in Chapter 1 and 7 in Chapter 3, 4.3.2, 4.3.7, 4.4.1, 4.5.2, Table 2.4}

D. Strengthening the Global Response in the Context of Sustainable Development and Efforts to Eradicate Poverty

D1. Estimates of the global emissions outcome of current nationally stated

mitigation ambitions as submitted under the Paris Agreement would lead to global greenhouse gas emissions[18] in 2030 of 52–58 GtCO$_2$eq yr^{-1} (*medium confidence*). Pathways reflecting these ambitions would not limit global warming to 1.5°C, even if supplemented by very challenging increases in the scale and ambition of emissions reductions after 2030 (*high confidence*). Avoiding overshoot and reliance on future large-scale deployment of carbon dioxide removal (CDR) can only be achieved if global CO$_2$ emissions start to decline well before 2030 (*high confidence*). {1.2, 2.3, 3.3, 3.4, 4.2, 4.4, Cross-Chapter Box 11 in Chapter 4}

D1.1. Pathways that limit global warming to 1.5°C with no or limited overshoot show clear emission reductions by 2030 (*high confidence*). All but one show a decline in global greenhouse gas emissions to below 35 GtCO$_2$eq yr^{-1} in 2030, and half of available pathways fall within the 25–30 GtCO$_2$eq yr^{-1} range (interquartile range), a 40–50% reduction from 2010 levels (*high confidence*). Pathways reflecting current nationally stated mitigation ambition until 2030 are broadly consistent with cost-effective pathways that result in a global warming of about 3°C by 2100, with warming continuing afterwards (*medium* *confidence*). {2.3.3, 2.3.5, Cross-Chapter Box 11 in Chapter 4, 5.5.3.2}

D1.2. Overshoot trajectories result in higher impacts and associated challenges compared to pathways that limit global warming to 1.5°C with no or limited overshoot (*high confidence*). Reversing warming after an overshoot of 0.2°C or larger during this century would require upscaling and deployment of CDR at rates and volumes that might not be achievable given considerable implementation challenges (*medium confidence*). {1.3.3, 2.3.4, 2.3.5, 2.5.1, 3.3, 4.3.7, Cross-Chapter Box 8 in Chapter 3, Cross-Chapter Box 11 in Chapter 4}

D1.3. The lower the emissions in 2030, the lower the challenge in limiting global warming to 1.5°C after 2030 with no or limited overshoot (*high confidence*). The challenges from delayed actions to reduce greenhouse gas emissions include the risk of cost escalation, lock-in in carbon-emitting infrastructure, stranded assets, and reduced flexibility in future response options in the medium to long-term (*high confidence*). These may increase uneven distributional impacts between countries at different stages of development (*medium confidence*). {2.3.5, 4.4.5, 5.4.2}

D2. The avoided climate change impacts on sustainable development, eradication of poverty and reducing inequalities would be greater if global warming were limited to 1.5°C rather than 2°C, if mitigation and adaptation synergies are maximized while trade-offs are minimized (*high confidence*). {1.1, 1.4, 2.5, 3.3, 3.4, 5.2, Table 5.1}

[18] GHG emissions have been aggregated with 100-year GWP values as introduced in the IPCC Second Assessment Report

D2.1. Climate change impacts and responses are closely linked to sustainable development which balances social well-being, economic prosperity and environmental protection. The United Nations Sustainable Development Goals (SDGs), adopted in 2015, provide an established framework for assessing the links between global warming of 1.5°C or 2°C and development goals that include poverty eradication, reducing inequalities, and climate action (*high confidence*) {Cross-Chapter Box 4 in Chapter 1, 1.4, 5.1}

D2.2. The consideration of ethics and equity can help address the uneven distribution of adverse impacts associated with 1.5°C and higher levels of global warming, as well as those from mitigation and adaptation, particularly for poor and disadvantaged populations, in all societies (*high confidence*). {1.1.1, 1.1.2, 1.4.3, 2.5.3, 3.4.10, 5.1, 5.2, 5.3. 5.4, Cross-Chapter Box 4 in Chapter 1, Cross-Chapter Boxes 6 and 8 in Chapter 3, and Cross-Chapter Box 12 in Chapter 5}

D2.3. Mitigation and adaptation consistent with limiting global warming to 1.5°C are underpinned by enabling conditions, assessed in SR1.5 across the geophysical, environmental-ecological, technological, economic, socio-cultural and institutional dimensions of feasibility. Strengthened multi-level governance, institutional capacity, policy instruments, technological innovation and transfer and mobilization of finance, and changes in human behaviour and lifestyles are enabling conditions that enhance the feasibility of mitigation and adaptation options for 1.5°C consistent systems transitions. *(high confidence)* {1.4, Cross-Chapter Box 3 in Chapter 1, 4.4, 4.5, 5.6}

D3. Adaptation options specific to national contexts, if carefully selected together with enabling conditions, will have benefits for sustainable development and poverty reduction with global

warming of 1.5°C, although trade-offs are possible (*high confidence*). {1.4, 4.3, 4.5}

D3.1. Adaptation options that reduce the vulnerability of human and natural systems have many synergies with sustainable development, if well managed, such as ensuring food and water security, reducing disaster risks, improving health conditions, maintaining ecosystem services and reducing poverty and inequality (*high confidence*). Increasing investment in physical and social infrastructure is a key enabling condition to enhance the resilience and the adaptive capacities of societies. These benefits can occur in most regions with adaptation to 1.5°C of global warming (*high confidence*). {1.4.3, 4.2.2, 4.3.1, 4.3.2, 4.3.3, 4.3.5, 4.4.1, 4.4.3, 4.5.3, 5.3.1, 5.3.2}

D3.2. Adaptation to 1.5°C global warming can also result in trade–offs or maladaptations with adverse impacts for sustainable development. For example, if poorly designed or implemented, adaptation projects in a range of sectors can increase greenhouse gas emissions and water use, increase gender and social inequality, undermine health conditions, and encroach on natural ecosystems (*high confidence*). These trade-offs can be reduced by adaptations that include attention to poverty and sustainable development (*high* *confidence*). {4.3.2, 4.3.3, 4.5.4, 5.3.2; Cross-Chapter Boxes 6 and 7 in Chapter 3}

D3.3. A mix of adaptation and mitigation options to limit global warming to 1.5°C, implemented in a participatory and integrated manner, can enable rapid, systemic transitions in urban and rural areas (*high confidence*). These are most effective when aligned with economic and sustainable development, and when local and regional governments and decision makers are supported by national governments (*medium confidence*) {4.3.2, 4.3.3, 4.4.1, 4.4.2}

D3.4. Adaptation options that also mitigate emissions can provide synergies and cost savings in most sectors and system transitions, such as when land management reduces emissions and disaster risk, or when low carbon buildings are also designed for efficient cooling. Trade-offs between mitigation and adaptation, when limiting global warming to 1.5°C, such as when bioenergy crops, reforestation or afforestation encroach on land needed for agricultural adaptation, can undermine food security, livelihoods, ecosystem functions and services and other aspects of sustainable development. (*high confidence*) {3.4.3, 4.3.2, 4.3.4, 4.4.1, 4.5.2, 4.5.3, 4.5.4}

D4. Mitigation options consistent with 1.5°C pathways are associated with multiple synergies and trade-offs across the Sustainable Development Goals (SDGs). While the total number of possible synergies exceeds the number of trade-offs, their net effect will depend on the pace and magnitude of changes, the composition of the mitigation portfolio and the management of the transition. *(high confidence)* **(Figure SPM.4) {2.5, 4.5, 5.4}**

D4.1. 1.5°C pathways have robust synergies particularly for the SDGs 3 (health), 7 (clean energy), 11 (cities and communities), 12 (responsible consumption and production), and 14 (oceans) (*very high confidence*). Some 1.5°C pathways show potential trade-offs with mitigation for SDGs 1 (poverty), 2 (hunger), 6 (water), and 7 (energy access), if not carefully managed (*high confidence*) (Figure SPM.4). {5.4.2; Figure 5.4, Cross-Chapter Boxes 7 and 8 in Chapter 3}

D4.2. 1.5°C pathways that include low energy demand (e.g., see P1 in Figure SPM.3a and SPM.3b), low material consumption, and low GHG-intensive food consumption have the most pronounced synergies and the lowest number of trade-offs with respect to sustainable development and the SDGs (*high confidence*). Such pathways would reduce dependence on CDR. In modelled pathways sustainable development, eradicating poverty and reducing inequality can support limiting warming to 1.5°C. (*high confidence*) (Figure SPM.3b, Figure SPM.4) {2.4.3, 2.5.1, 2.5.3, Figure 2.4, Figure 2.28, 5.4.1, 5.4.2, Figure 5.4}

D4.3. 1.5°C and 2°C modelled pathways often rely on the deployment of large-scale land-related measures like afforestation and bioenergy supply, which, if poorly managed, can compete with food production and hence raise food security concerns (*high confidence*). The impacts of carbon dioxide removal (CDR) options on SDGs depend on the type of options and the scale of deployment (*high confidence*). If poorly implemented, CDR options such as BECCS and AFOLU options would lead to trade-offs. Context-relevant design and implementation requires considering people's needs, biodiversity, and other sustainable development dimensions (*very high confidence*). {Figure SPM.4, 5.4.1.3, Cross-Chapter Box 7 in Chapter 3}

D4.4. Mitigation consistent with 1.5°C pathways creates risks for sustainable development in regions with high dependency on fossil fuels for revenue and employment generation (*high confidence*).

Policies that promote diversification of the economy and the energy sector can address the associated challenges (*high confidence*). {5.4.1.2, Box 5.2}

D4.5. Redistributive policies across sectors and populations that shield the poor and vulnerable can resolve trade-offs for a range of SDGs, particularly hunger, poverty and energy access. Investment needs for such complementary policies are only a small fraction of the overall mitigation investments in 1.5°C pathways. (*high confidence*) {2.4.3, 5.4.2, Figure 5.5}

Indicative linkages between mitigation options and sustainable development using SDGs (The linkages do not show costs and benefits)

Mitigation options deployed in each sector can be associated with potential positive effects (synergies) or negative effects (trade-offs) with the Sustainable Development Goals (SDGs). The degree to which this potential is realized will depend on the selected portfolio of mitigation options, mitigation policy design, and local circumstances and context. Particularly in the energy-demand sector, the potential for synergies is larger than for trade-offs. The bars group individually assessed options by level of confidence and take into account the relative strength of the assessed mitigation-SDG connections.

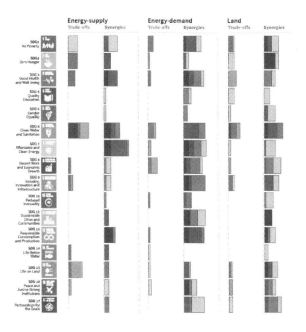

Figure SPM.4: Potential synergies and trade-offs between the sectoral portfolio of climate change mitigation options and the Sustainable Development Goals (SDGs). The SDGs serve as an analytical framework for the assessment of the different sustainable development dimensions, which extend beyond the time frame of the 2030

SDG targets. The assessment is based on literature on mitigation options that are considered relevant for 1.5oC. The assessed strength of the SDG interactions is based on the qualitative and quantitative assessment of individual mitigation options listed in Table 5.2. For each mitigation option, the strength of the SDG-connection as well as the associated confidence of the underlying literature (shades of green and red) was assessed. The strength of positive connections (synergies) and negative connections (trade-offs) across all individual options within a sector (see Table 5.2) are aggregated into sectoral potentials for the whole mitigation portfolio. The (white) areas outside the bars, which indicate no interactions, have *low confidence* due to the uncertainty and limited number of studies exploring indirect effects. The strength of the connection considers only the effect of mitigation and does not include benefits of avoided impacts. SDG 13 (climate action) is not listed because mitigation is being considered in terms of interactions with SDGs and not vice versa. The bars denote the strength of the connection, and do not consider the strength of the impact on the SDGs. The energy demand sector comprises behavioural responses, fuel switching and efficiency options in the transport, industry and building sector as well as carbon capture options in the industry sector. Options assessed in the energy supply sector comprise biomass and non-biomass renewables, nuclear, CCS with bio-energy, and CCS with fossil fuels. Options in the land sector comprise agricultural and forest options, sustainable diets & reduced food waste, soil sequestration, livestock & manure management, reduced deforestation, afforestation & reforestation, responsible sourcing. In addition to this figure, options in the ocean sector are discussed in the underlying report. {5.4, Table 5.2, Figure 5.2}

Statement for knowledge gap:

Information about the net impacts of mitigation on sustainable development in 1.5°C pathways is available only for a limited number of SDGs and mitigation options. Only a limited number of studies have assessed the benefits of avoided climate change impacts of 1.5°C pathways for the SDGs, and the co-effects of adaptation for mitigation and the SDGs. The assessment of the indicative mitigation potentials in Figure SPM.4 is a step further from AR5 towards a more comprehensive and integrated assessment in the future.

D5. Limiting the risks from global warming of 1.5°C in the context of sustainable development and poverty

eradication implies system transitions that can be enabled by an increase of adaptation and mitigation investments, policy instruments, the acceleration of technological innovation and behaviour changes (*high confidence*). {2.3, 2.4, 2.5, 3.2, 4.2, 4.4, 4.5, 5.2, 5.5, 5.6}

D5.1. Directing finance towards investment in infrastructure for mitigation and adaptation could provide additional resources. This could involve the mobilization of private funds by institutional investors, asset managers and development or investment banks, as well as the provision of public funds. Government policies that lower the risk of low-emission and adaptation investments can facilitate the mobilization of private funds and enhance the effectiveness of other public policies. Studies indicate a number of challenges including access to finance and mobilisation of funds (*high confidence*) {2.5.2, 4.4.5}

D5.2. Adaptation finance consistent with global warming of 1.5°C is difficult to quantify and compare with 2°C. Knowledge gaps include insufficient data to calculate specific climate resilience-enhancing investments, from the provision of currently underinvested basic infrastructure. Estimates of the costs of adaptation might be lower at global warming of 1.5°C than for 2°C.

Adaptation needs have typically been supported by public sector sources such as national and subnational government budgets, and in developing countries together with support from development assistance, multilateral development banks, and UNFCCC channels (*medium confidence*). More recently there is a growing understanding of the scale and increase in NGO and private funding in some regions (*medium confidence*). Barriers include the scale of adaptation financing, limited capacity and access to adaptation finance (*medium confidence*). {4.4.5, 4.6}

D5.3. Global model pathways limiting global warming to 1.5°C are projected to involve the annual average investment needs in the energy system of around 2.4 trillion USD2010 between 2016 and 2035 representing about 2.5% of the world GDP (*medium confidence*). {2.5.2, 4.4.5, Box 4.8}

D5.4. Policy tools can help mobilise incremental resources, including through shifting global investments and savings and through market and non-market based instruments as well as accompanying measures to secure the equity of the transition, acknowledging the challenges related with implementation including those of energy costs, depreciation of assets and impacts on international competition, and utilizing the opportunities to maximize co-benefits (*high confidence*) {1.3.3, 2.3.4, 2.3.5, 2.5.1, 2.5.2, Cross-Chapter Box 8 in Chapter 3 and 11 in Chapter 4, 4.4.5, 5.5.2}

D5.5. The systems transitions consistent with adapting to and limiting global warming to 1.5°C include the widespread adoption of new and possibly disruptive technologies and practices and enhanced climate-driven innovation. These imply enhanced technological innovation capabilities, including in industry and finance. Both national innovation policies and international cooperation can contribute to the development, commercialization and widespread adoption of mitigation and adaptation technologies. Innovation policies may be more effective when they combine public support for research and development with policy mixes that provide incentives for technology diffusion. (*high confidence*) {4.4.4, 4.4.5}.

D5.6. Education, information, and community approaches, including those that are informed by Indigenous knowledge and local knowledge, can accelerate the wide scale behaviour changes consistent with adapting to and limiting global warming to 1.5°C. These approaches are more effective when combined with other policies and tailored to the motivations, capabilities, and resources of specific actors and contexts (*high confidence*). Public acceptability can enable or inhibit the implementation of policies and measures to limit global warming to 1.5°C and to adapt to the consequences. Public acceptability depends on the individual's evaluation of expected policy consequences, the perceived fairness of the distribution of these consequences, and perceived fairness of decision procedures (*high confidence*). {1.1, 1.5, 4.3.5, 4.4.1, 4.4.3, Box 4.3, 5.5.3, 5.6.5}

D6. Sustainable development supports, and often enables, the fundamental societal and systems transitions and transformations that help limit global warming to 1.5°C. Such changes facilitate the pursuit of climate-resilient development pathways that achieve ambitious mitigation and adaptation in conjunction with poverty eradication and efforts to reduce inequalities *(high confidence)*. {Box 1.1, 1.4.3, Figure 5.1, 5.5.3, Box 5.3}

D6.1. Social justice and equity are core aspects of climate-resilient development pathways that aim to limit global warming to 1.5°C as they address challenges and inevitable trade-offs, widen opportunities, and ensure that options, visions, and values are deliberated, between and within countries and communities, without making the poor and disadvantaged worse off (*high confidence*). {5.5.2, 5.5.3, Box 5.3, Figure 5.1, Figure 5.6, Cross-Chapter Boxes 12 and 13 in Chapter 5}

D6.2. The potential for climate-resilient development pathways differs between and within regions and nations, due to different development contexts and systemic vulnerabilities (*very high confidence*). Efforts along such pathways to date have been limited (*medium confidence*) and enhanced efforts would involve strengthened and timely action from all countries and non-state actors (*high confidence*). {5.5.1, 5.5.3, Figure 5.1}

D6.3. Pathways that are consistent with sustainable development show fewer mitigation and adaptation challenges and are associated with lower mitigation costs. The large majority of modelling studies could not construct pathways characterized by lack of international cooperation, inequality and poverty that were able to limit global warming to 1.5°C. (*high confidence*) {2.3.1, 2.5.3, 5.5.2}

D7. Strengthening the capacities for climate action of national and sub-

national authorities, civil society, the private sector, indigenous peoples and local communities can support the implementation of ambitious actions implied by limiting global warming to 1.5°C (*high confidence*). International cooperation can provide an enabling environment for this to be achieved in all countries and for all people, in the context of sustainable development. International cooperation is a critical enabler for developing countries and vulnerable regions (*high confidence*). {1.4, 2.3, 2.5, 4.2, 4.4, 4.5, 5.3, 5.4, 5.5, 5.6, 5, Box 4.1, Box 4.2, Box 4.7, Box 5.3, Cross-Chapter Box 9 in Chapter 4, Cross-Chapter Box 13 in Chapter 5}

D7.1. Partnerships involving non-state public and private actors, institutional investors, the banking system, civil society and scientific institutions would facilitate actions and responses consistent with limiting global warming to 1.5°C (*very high confidence*). {1.4, 4.4.1, 4.2.2, 4.4.3, 4.4.5, 4.5.3, 5.4.1, 5.6.2, Box 5.3}.

D7.2. Cooperation on strengthened accountable multilevel governance that includes non-state actors such as industry, civil society and scientific institutions, coordinated sectoral and cross-sectoral policies at various governance levels, gender-sensitive policies, finance including innovative financing and cooperation on technology development and transfer can ensure participation, transparency, capacity building, and learning among different players (*high confidence*). {2.5.2, 4.2.2, 4.4.1, 4.4.2, 4.4.3, 4.4.4, 4.5.3, Cross-Chapter Box 9 in Chapter 4, 5.3.1, 4.4.5, 5.5.3, Cross- Chapter Box 13 in Chapter 5, 5.6.1, 5.6.3}

D7.3. International cooperation is a critical enabler for developing countries and vulnerable regions to strengthen their action for the implementation of 1.5°C-consistent climate responses, including through enhancing access to finance and technology and enhancing domestic capacities, taking into account national and local circumstances and needs (*high confidence*). {2.3.1, 4.4.1, 4.4.2, 4.4.4, 4.4.5, 5.4.1 5.5.3, 5.6.1, Box 4.1, Box 4.2, Box 4.7}.

D7.4. Collective efforts at all levels, in ways that reflect different circumstances and capabilities, in the pursuit of limiting global warming to 1.5°C, taking into account equity as well as effectiveness, can facilitate strengthening the global response to climate change, achieving sustainable development and eradicating poverty (*high confidence*). {1.4.2, 2.3.1, 2.5.2, 4.2.2, 4.4.1, 4.4.2, 4.4.3,

4.4.4, 4.4.5, 4.5.3, 5.3.1, 5.4.1, 5.5.3, 5.6.1, 5.6.2, 5.6.3}

Box SPM 1: Core Concepts Central to this Special Report

Global mean surface temperature (GMST): Estimated global average of near-surface air temperatures over land and sea-ice, and sea surface temperatures over ice-free ocean regions, with changes normally expressed as departures from a value over a specified reference period.
When estimating changes in GMST, near-surface air temperature over both land and oceans are also used.[19] {1.2.1.1}

Pre-industrial: The multi-century period prior to the onset of large-scale industrial activity around 1750. The reference period 1850–1900 is used to approximate pre-industrial GMST. {1.2.1.2}

Global warming: The estimated increase in GMST averaged over a 30-year period, or the 30-year period centered on a particular year or decade, expressed relative to pre-industrial levels unless otherwise specified. For 30-year periods that span past and future years, the current multi-decadal warming trend is assumed to continue. {1.2.1}

Net zero CO$_2$ emissions: Net-zero carbon dioxide (CO$_2$) emissions are achieved when anthropogenic CO$_2$ emissions are balanced globally by anthropogenic CO$_2$ removals over a specified period.

Carbon dioxide removal (CDR): Anthropogenic activities removing CO$_2$ from the atmosphere and durably storing it in geological, terrestrial, or ocean reservoirs, or in products. It includes existing and potential anthropogenic enhancement of biological or geochemical sinks and direct air capture and storage, but excludes natural CO$_2$ uptake not directly caused by human activities.

Total carbon budget: Estimated cumulative net global anthropogenic CO$_2$ emissions from the preindustrial period to the time that anthropogenic CO$_2$ emissions reach net zero that would result, at some probability, in limiting global warming to a given level, accounting for the impact of other anthropogenic emissions. {2.2.2}

Remaining carbon budget: Estimated cumulative net global anthropogenic CO$_2$ emissions from a given start date to the time that anthropogenic CO$_2$ emissions reach net zero that would result, at some probability, in limiting global warming to a given level, accounting for the impact of other anthropogenic emissions. {2.2.2}

Temperature overshoot: The temporary exceedance of a specified level of global warming.

Emission pathways: In this Summary for Policymakers, the modelled trajectories of global anthropogenic emissions over the 21st century are termed emission pathways. Emission pathways are classified by their temperature trajectory over the 21st century: pathways giving at least 50% probability based on current knowledge of limiting global warming to below 1.5°C are classified as 'no overshoot'; those limiting warming to below 1.6°C and returning to 1.5°C by 2100 are classified as '1.5°C limited-overshoot'; while those exceeding 1.6°C but still returning to 1.5°C by 2100 are classified as 'higher-overshoot'.

---

[19] Past IPCC reports, reflecting the literature, have used a variety of approximately equivalent metrics of GMST change.

Impacts: Effects of climate change on human and natural systems. Impacts can have beneficial or adverse outcomes for livelihoods, health and well-being, ecosystems and species, services, infrastructure, and economic, social and cultural assets.

Risk: The potential for adverse consequences from a climate-related hazard for human and natural systems, resulting from the interactions between the hazard and the vulnerability and exposure of the affected system. Risk integrates the likelihood of exposure to a hazard and the magnitude of its impact. Risk also can describe the potential for adverse consequences of adaptation or mitigation responses to climate change.

Climate-resilient development pathways (CRDPs): Trajectories that strengthen sustainable development at multiple scales and efforts to eradicate poverty through equitable societal and systems transitions and transformations while reducing the threat of climate change through ambitious mitigation, adaptation, and climate resilience.

[1] *New York Times, March 3, 2018*
[2] Ellis, Erle C. 2018. *Anthropocene: A Very Short Introduction (Very Short Introductions)* (p. 53). OUP Oxford. Kindle Edition. Graphs adapted from *Global Change and the Earth System* by W. Steffen, A. Sanderson, P. D . Tyson, J. Jäger, P. A . Matson, B. Moore III, F. Oldfield, K. Richardson, H. J. Schellnhuber, B. L. Turner II, R. J. Wasson, 2006.
[3] BBC.com, December 3, 2018
[4] McPherson, William. 1973. *Ideology and Change: Radicalism and Fundamentalism in America.* National Press Books.
[5] McPherson, William. 2012. *Ideology and War.* Kindle Edition
[6] McPherson, William. 2014. *Ideology versus Science.* Amazon.
McPherson, William. 2015. *Climate, Weather and Ideology.* Amazon.
McPherson, William. 2016. *Sabotaging the Planet.* Amazon.
[7] CNNMoney, August 17, 2018
[8] Bastardi, Joe. 2018. *The Climate Chronicles: Inconvenient Revelations You Won't Hear From Al Gore--And Others.* An Original Publication of Relentless Thunder Press. Kindle Edition.
[9] Incropera, Frank P.. 2017. *Climate Change: A Wicked Problem: Complexity and Uncertainty at the Intersection of*

Science, Economics, Politics, and Human Behavior (pp. 146-147). Cambridge University Press. Kindle Edition.

[10] "The Climate-Wrecking Industry… and How to Beat It," *The Nation,* August 30, 2018

[11] "Climate Uncovered: Media Fail to Connect Hurricane Florence to Climate Change", *Public Citizen,* September 19, 2018

[12] One oil executive poured cold water on that goal: "Even with large advances in renewable energy, [Shell CEO Ben van Beurden] said, the share of world energy met by oil and gas would decline from 85 percent to 75 percent by the middle of the century, a time when the IPCC said net carbon dioxide emissions should drop to zero." *Washington Post,* December 4, 2018

[13] https://wedocs.unep.org/bitstream/handle/20.500.11822/26879/EGR2018_ESEN.pdf?sequence=10

[14] *Guardian*, October 31, 2018

[15] *New York Times,* December 29, 2018

[16] Monbiot, George. 2018. *Out of the Wreckage: A New Politics for an Age of Crisis* (Kindle Locations 1699-1703). Verso Books. Kindle Edition

[17] *New York Times* Magazine, September 13, 1970

[18] *New York Times,* August 4, 2018

[19] https://www.cruz.senate.gov/?p=press_rele

ase&id=3904. Emphasis added.

[20] Singer, Fred, "The Sea Is Rising, but Not Because of Climate Change," *Wall Street Journal,* May 15, 2018

[21] *Climate Liability Project lead for Greenpeace,* Huffington Post, March 24, 2018

[22] *New York Times,* November 19, 2018

[23] Lewin, Bernie. 2017. *Searching for the Catastrophe Signal: The Origins of the Intergovernmental Panel on Climate Change* . Global Warming Policy Foundation.

[24] Morano, Marc. 2018. The Politically Incorrect Guide to Climate Change (The Politically Incorrect Guides). Regnery Publishing. Kindle Edition.

[25] NOAA: National Oceanic and Atmospheric Administration data, 2018

[26] Michael White, "Resistance to Climate Change Is Killing the Government's Ability to Use Science," *Pacific Standard,* May 23, 2014

[27] https://envirodatagov.org/website-monitoring/

[28] https://www.nhtsa.gov/sites/nhtsa.dot.gov/files/documents/ld_cafe_my202126_deis_0.pdf

[29] *New York Times,* December 29, 2018

[30] *Scientific American,* October 2, 2018

[31] Methane is about 25 times more potent than carbon dioxide at trapping heat.

[32] *New York Times,* September 10, 2018
[33] *New York Times,* February 7, 2017
[34] Incropera, Frank P.. 2017. *Climate Change: A Wicked Problem: Complexity and Uncertainty at the Intersection of Science, Economics, Politics, and Human Behavior.* Cambridge University Press. Kindle Edition
[35] Quirk, Joe. 2018. *Seasteading: How Floating Nations Will Restore the Environment, Enrich the Poor, Cure the Sick, and Liberate Humanity from Politicians* (pp. 69-70). Free Press. Kindle Edition.
[36] Some collapse ideologues will admit this, but fall back on an argument that wind and solar are unreliable and a danger to the electrical grid. See Rogers, Norman. 2018. *Dumb* Energy*: A Critique of Wind and Solar Energy* (p. 34). Dumb Energy Publishing. Kindle Edition
[37] Perry, Rick. 2010. *Fed Up! Our Fight to Save America from Washington.* Little, Brown
[38] *New* York *Times,* June 19, 2018
[39] *New York Times,* September 7, 2018
[40] "Climate confusion among U.S. teachers," *Science,* 12 Feb 2016: "Climate Change and Energy Issues, Pew Research Center, July 1, 2015
[41] https://www.pdc.wa.gov/browse/campaign-

explorer/committee?filer_id=NO1631%205
07&election_year=2018

[42] Johansen, Bruce. 2015. *Eco-Hustle! Global Warming, Greenwashing, and Sustainability: Global Warming, Greenwashing, and Sustainability* (p. 229). ABC-CLIO. Kindle Edition.

[43] Ball, Tim. 2016. *Human Caused Global Warming.* Tellwell Talent. Kindle Edition. Despite Ball's misstatement, the hottest years on record are 2014, 2015, 2015 and 2017, all warmer than 1998 when he starts his trace of "decline." There were some cooler years in between but the overall trend is up. This was one of the most persistent canards in the denial playbook, and it is amazing that it persists to this day.

[44] I discuss this more intensively in *Ideology versus Science.*

[45] Lifton, Robert Jay. *The Climate Swerve: Reflections on Mind, Hope, and Survival.* The New Press. Kindle Edition

[46] *Chicago Tribune,* July 27, 2018

[47] *The Week*, September 15, 2018

[48] *New York Times,* November 26, 2018

[49] https://nca2018.globalchange.gov/

[50] *Politico,* November 28, 2018

[51] https://t.co/aTkrBNh9IL

[52] *New York Times,* July 29, 2018

[53] "Japan is the Latest Country to Break a Heat Record," by Brian Kahn, *Earther,* July 23, 2018.

[54] *Los* Angeles *Times*, July 31, 2018
[55] McPherson, William. 2015, *Climate, Weather and Ideology.* Amazon
[56] A partial list:
> Morano, Marc. 2018. *The Politically Incorrect Guide to Climate Change.* Thomas Woods
> Tisdale, Bob. 2018. *Dad, Why Are You A Global Warming Denier?: A Short Story That's Right for the Times.* Kindle Edition.
> Darwall, Rupert. 2017. *Green Tyranny.* Encounter Books.
> King, M. 2017. *Climate Bogeyman: The* CriNew minal *Insanity of the Global Warming / Climate Change Hoax.* Kindle Edition.
> Madden, Jack. 2017. *Inconvenient Facts: proving Global Warming is a Hoax.* Kindle Edition.

[57] Andreas Malm. 2016. *Fossil Capital: The Rise of Steam Power and the Roots of Global Warming.* Verso.
[58] Lifton, Robert Jay. 2018. *The Climate Swerve: Reflections on Mind, Hope, and Survival.* The New Press. Kindle Edition
[59] Scientific *American,* October 4, 2018
[60] AP, December 10, 2018
[61] Incropera, Frank P.. 2017. *Climate Change: A Wicked Problem: Complexity and Uncertainty at the Intersection of Science, Economics, Politics, and Human*

Behavior.
[62] Nesbit, Jeff. 2018. *This Is the Way the World Ends: How Droughts and Die-offs, Heat Waves and Hurricanes Are Converging on America.* St. Martin's Press. Kindle Edition.
[63] *New York Times,* December 10, 2018
[64] Bastardi, Joe. 2018. *The Climate Chronicles: Inconvenient Revelations You Won't Hear From Al Gore--And Others.* An Original Publication of Relentless Thunder Press. Kindle Edition.
[65] *New York Times,* August 30, 2018
[66] Davies, Kate. 2013. *The Rise of the U.S. Environmental Health Movement* (p. 67). Rowman & Littlefield Publishers. Kindle Edition.
[67] *New York Times,* August 24, 2018
[68] *Guardian,* August 13, 2018
[69] Hochschild, Arlie Russell. 2018. *Strangers in Their Own Land: Anger and Mourning on the American Right* (p. 53). The New Press. Kindle Edition
[70] McPherson, 2014, 2015. The first is a quote from Illinois congressman John Shimkus, and the second is from Oklahoma Senator James Inhofe.
[71] Vollman, William. 2018. *No Good Alternative.* Penguin.
[72] Angus, Ian. 2018. *Facing the Anthropocene: Fossil Capitalism and the Crisis of the Earth System* (Kindle Locations

907-908). Monthly Review Press. Kindle Edition.

[73] Bastardi, Joe. 2018. *The Climate Chronicles: Inconvenient Revelations You Won't Hear From Al Gore--And Others*. An Original Publication of Relentless Thunder Press. Kindle Edition.

[74] McPherson, 2015

[75] Morano, Marc. 2018. *The Politically Incorrect Guide to Climate Change* (The Politically Incorrect Guides). Regnery Publishing. Kindle Edition.

[76] Wishart, Ian. 2013. *Totalitaria: What If The Enemy Is The State?*. Howling At The Moon Publishing Ltd.

[77] Klein, Naomi. 2017. *No Is Not Enough: Resisting Trump's Shock Politics and Winning the World We Need* (p. 66). Haymarket Books. Kindle Edition.

[78] Klein, Naomi. 2017. *No Is Not Enough: Resisting Trump's Shock Politics and Winning the World We Need* (p. 66). Haymarket Books. Kindle Edition.

[79] *New York Times,* November 24, 2018

[80] *New York Times,* December 6, 2018

[81] *New York Times,* December 15, 2018

[82] *New York Times,* December 20, 2018. "Yellow Vest" refers to the safety vests required by French law for drivers, and the protestors used these vests to build solidarity.

[83] Malm, *ibid.*

84 *New York Times,* August 2, 2018
85 Quoted *in New York Times,* August 7, 2018
86 *New York Times,* August 3, 2018
87 *New York Times,* August 23, 2018
88 McClatchy Washington Bureau, August 7, 2018
89 *Business Insider,* October 10, 2018
90 *E&E News,* November 5, 2018. For an extensive description of Heartland Institute reports, see *Ideology versus Science.*
91 Twitter, November 21, 2018
92 *New York Times,* October 10, 2018
93 Vinod, Thomas. 2017. *Climate Change and Natural Disasters.* Routledge.
94 Harari, Yuval Noah. 2018. *21 Lessons for the 21st Century.* (p. 238). Random House Publishing Group. Kindle Edition. Emphasis added.
95 Joel Wainwright; Geoff Mann. 2018. *Climate Leviathan: A Political Theory of Our Planetary Future.* Verso.
96 Lifton, Robert Jay. 2018. *The Climate Swerve: Reflections on Mind, Hope, and Survival.* The New Press. Kindle Edition.
97 Attributed to Rep. John Shimkus, R-IL
98 *New York Times,* December 5, 2018
99 *Guardian,* October 30, 2018.
100 Nesbit, Jeff. 2018. *This Is the Way the World Ends: How Droughts and Die-offs, Heat Waves and Hurricanes Are Converging on America* (Kindle Locations

3963-3967). St. Martin's Press. Kindle Edition.

[101] Why do those states have 40% of the economic activity but only 35% of the emissions of the U.S.? Many have already implemented climate policies that lower their emissions relative to other states in the United States, so their emissions per dollar of economic activity are less than the rest. See report at https://www.bbhub.io/dotorg/sites/28/2018/09/Fulfilling-Americas-Pledge-2018.pdf

[102] *Foreign Affairs,* September 24, 2018

[103] Davies, Kate. 2013. *The Rise of the U.S. Environmental Health Movement* (p. 97). Rowman & Littlefield Publishers. Kindle Edition.

[104] *U.S. News,* September 17, 2018

[105] Twitter, December 5, 2018

[106] For the full text of the Paris Agreement, see *Sabotaging the Planet,* Amazon

[107] https://unfccc.int/resource/docs/2015/cop21/eng/10a01.pdf

[108] *The Outline,* November 18, 2018

[109] Joel Wainwright; Geoff Mann. *2016. Climate Leviathan: A Political Theory of Our Planetary Future.* Verso.

[110] Politifact, June 1, 2017. These data were derived from a March 2017 study prepared by NERA Economic Consulting (the acronym is not spelled out). NERA analyzed the potential impact of *hypothetical*

regulatory actions necessary to meet the goals of the Paris Agreement. Their analyses have been questioned by a number of leading economists and statisticians. NERA is viewed as a captive of the denial wing of conservative organizations.

[111] CBS, June 2, 2017

[112] *Forbes*, January 27, 2017; insideclimatenews.org, May 27, 2017

[113] The *Stern* Review on the Economics of *Climate Change,* Government of the United Kingdom, 30 October 2006 by economist Nicholas *Stern*,

[114] Mishra-Marzetti, Manish. 2018. *Justice on Earth: People of Faith Working at the Intersections of Race, Class, and the Environment.* Skinner House Books. Kindle Edition.

[115] Mishra-Marzetti, Manish. 2018. *Justice on Earth: People of Faith Working at the Intersections of Race, Class, and the Environment.* Skinner House Books. Kindle Edition.

[116] Trump's approach to many policies is based on fear. See Woodward, Bob. 2018. *Fear: Trump on the White House.* Simon and Schuster.

[117] Based on a MIT Study

[118] CBS, June 3, 2017. Trump cited an MIT research paper in making this statement. The co-founder of the MIT program on climate change, Jake Jacoby,

says that the Trump administration cited an outdated report, taken out of context. Jacoby rebutted him with data showing the actual global impact of meeting targets under the Paris Accord would be to curb rising temperatures by 1 degree Celsius, or 1.8 degrees Fahrenheit.

[119] Anderson, Kevin, and Alice Bows. 2010. "A 2C Target? Get Real, Because 4C is on its Way." *Parliamentary Brief.*
[120] *Japan Times,* September 30, 2018
[121] Guardian, May 11, 2017
[122] Source: NOAA
[123] https://www.climate.gov/news-features/understanding-climate/climate-change-ocean-heat-content
[124] "A Congressional Hearing on Climate Change Turned Into a Circus of Absurdist Climate Denial," Carly Cassella, Science As Fact, *Alert,* 18 May 2018
[125] Bloomberg New Energy, June, 2017
[126] *New York Times*, August 24, 2018
[127] *New York Times*, August 21, 2018
[128] "Cambridge Analytica, Facebook and the climate change angle," Greenbiz, March 26, 2018
[129] Harari, Yuval Noah. 2018. *21 Lessons for the 21st Century.* (p. 244). Random House Publishing Group. Kindle Edition.
[130] *New York Times,* August 15, 2018
[131] *Climate Home News,* June 10, 2018
[132] "Mike Pompeo, Climate Policy Foe,

Picked to Replace Tillerson as Secretary of State," *Inside Climate News,* March 13, 2018

[133] *Bloomberg Politics*, March 15, 2018

[134] McPherson, 2016

[135] *New York Times,* August 4, 2018

[136] "Every climate denier in Trump's cabinet: Trump has surrounded himself with climate science deniers," Ryan Koronowski, Claire Moser, *ThinkProgress,* Feb 1, 2018:

- EPA Administrator Scott Pruitt: "I would not agree that [human activity is] a primary contributor to the global warming that we see."
- Energy Secretary Rick Perry: On whether carbon dioxide is the primary contributor to climate change, Perry stated that "no, most likely the primary control knob is the ocean waters and this environment that we live in."
- Interior Secretary Ryan Zinke: "It's not a hoax, but it's not proven science either."
- Vice President Mike Pence: "I don't know that that is a resolved issue in science today…"
- Attorney General Jeff Sessions: "…It's another to spend hundreds of billions of dollars each year to try to fight this global warming that we're not even sure

exists."
- OMB Director Mick Mulvaney: Asked "Is climate change driven by human-generated CO2 emissions a huge risk?" Mulvaney said, "I challenge the premise of your fact…"
- Director of National Intelligence Dan Coats: "There have always, in the history of the world, been reactions to different climate changes, and that is an issue that continues."
- Secretary of State Mike Pompeo: "Look, I think the science needs to continue to develop. There are scientists who think lots of different things about climate change. There's some who think we're warming, there's some who think we're cooling, there's some who think that the last 16 years have shown a pretty stable climate environment."
- Homeland Security Secretary Kirstjen Nielsen: At her Senate confirmation hearing, she said, "I do absolutely believe that the climate is changing," but that she is "not prepared to determine causation," following up that "she would review the science."
- Agriculture Secretary Sonny Perdue: "Whether temperatures are

unseasonably low or high, global warming is the culprit. Snowstorms, hurricanes, and tornadoes have been around since the beginning of time, but now they want us to accept that all of it is the result of climate change."
- HUD Secretary Ben Carson: Carson has said "there's always going to be either cooling or warming going on."
- Linda McMahon, administrator of the Small Business Administration: "I just don't think we have the answers as to why it changes… I'm not a scientist, so I couldn't pretend to understand all the reasons. But the bottom line is we really don't know."
- The only member of the cabinet who has never appeared to question the reality of climate change is Defense Secretary James Mattis.

[137] Klein, Naomi. 2017. *No Is Not Enough: Resisting Trump's Shock Politics and Winning the World We Need* (pp. 156-157). Haymarket Books. Kindle Edition.
[138] https://www.congress.gov/115/bills/hconres119/BILLS-115hconres119rfs.pdf
[139] "Pompeo, Trump and the Paris climate agreement," John Stossel, Fox News, March 20, 2018
[140] Morano, Marc. 2018. *The Politically Incorrect Guide to Climate Change (The*

Politically Incorrect Guides). Regnery Publishing. Kindle Edition.

[141] Ibid.

[142] "Alliance States Take the Lead: United States Climate Alliance: On Track for Paris and Thriving," *By Andrew M. Cuomo, Jerry Brown, and Jay Inslee [governors of New York, California and Washington, respectively),* 2017 Annual Report

[143] Ibid.

[144] *Huffpost,* November 10, 2018

[145] *New York Times,* September 18, 2018

[146] AGs' brief, filed May 30 2018 in U.S. District Court for the Southern District of New York. https://www.forbes.com/sites/legalnewsline/2018/06/04/states-against-climate-change-lawsuits-ask-for-dismissal-of-nycs/#1579329d97b8

[147] When both the Alliance members and a delegation of the U.S. government came to the 2017 COP, the government delegation was booed when it argued for continued development of fossil fuels. The Alliance delegation was greeted with open arms.

[148] KUOW.org, August 5, 2018

[149] Article 4 of the Paris Agreement reads: "In order to achieve the long-term temperature goal set out in Article 2, Parties aim to reach global peaking of greenhouse gas emissions as soon as possible, recognizing that peaking will take longer for

developing country Parties, and to undertake rapid reductions thereafter in accordance with best available science, so as to achieve a *balance between anthropogenic emissions by sources and removals by sinks of greenhouse gases in the second half of this century,* on the basis of equity, and in the context of sustainable development and efforts to eradicate poverty." (emphasis added) In other words, zero net emissions by 2050.

[150] Guardian, June 19, 2018
[151] McPherson, William. *Sabotaging the Planet: Denial and International Negotiations.* Kindle Edition.
[152] Morano, Marc. 2018. *The Politically Incorrect Guide to Climate Change (The Politically Incorrect Guides)* Regnery Publishing. Kindle Edition.
[153] Ibid.
[154] Xinhua 2018-07-07
[155] Johansen, Bruce. 2018. *Eco-Hustle! Global Warming, Greenwashing, and Sustainability: Global Warming, Greenwashing, and Sustainability* (p. 22). ABC-CLIO. Kindle Edition
[156] Mitchell, Timothy. 2011. *Carbon Democracy: Political Power in the Age of Oil* (p. 157). Verso Books. Kindle Edition.
[157] https://www.acesconnection.com/g/international-transformational-resilience-coalition-itrc/blog/article-shows-how-climate-change-

can-generate-harmful-psychosocial-maladies
[158] Lefebvre, Henri. 1992. *Production of Space*. Wiley-Blackwell.
[159] McKibben, Bill. 2014. *The End of Nature* (Second Edition). Random House.
[160] *UN News,* September 10, 2018
[161] AP, December 12, 2018
[162] Macy, Joanna. 2014. *Coming Back to Life* (p. 58). New Society Publishers. Kindle Edition.
[163] Martin Jr., Rod. 2018. *Climate Basics: Nothing to Fear* (Kindle Locations 168-169). Tharsis Highlands. Kindle Edition.
[164] "In November [2018], the number of electric vehicles in the United States hit the 1 million mark. But that was three years later than President Barack Obama's target, first issued in 2009. And that makes only a small dent in the nation's greenhouse gas emissions. Thanks to the growth in the car market, in 2016 there were nearly 12 million more cars with internal combustion engines emitting greenhouse gases than there were in 2008." *Washington Post,* December 4, 2018
[165] *New York Times, August 5, 2018,* emphasis added
[166] Ibid.
[167] https://www.channel3000.com, November 14, 2018
[168] CNN, December 4, 2018
[169] *Bloomberg Business Week,* October 8, 2018

[170] Ibid.
[171] Ibid.
[172] *New York Times,* August 5, 2018, emphasis added
[173] Ibid. Interestingly, George H.W. Bush did sign the UN Framework Convention on Climate Change, but Sununu probably was referring to agreeing on binding reduction commitments.
[174] Hamilton, Clive. 2018. *Defiant Earth: The Fate of Humans in the Anthropocene* (p. 148). Wiley. Kindle Edition.
[175] The IPCC, for example, always caveats its projections with a range of temperatures, such as 2C to 5C. See Appendix.
[176] *Washington Post,* November 4, 2018
[177] Kim Cobb, a professor of earth and atmospheric science at the Georgia Institute of Technology in Atlanta, quoted in the *New York Times,* August 10, 2018.
[178] Hamilton, Clive. 2018. *Defiant Earth: The Fate of Humans in the Anthropocene* (pp. 77-78). Wiley. Kindle Edition.
[179] *New York Times*, August 10, 2018
[180] Andreas Malm. 2016. *Fossil Capital: The Rise of Steam Power and the Roots of Global Warming.* Verso.
[181] Ozarko, Deb. 2018. *Beyond Hope: Letting Go of a World in Collapse* (p. 15). Deb Ozarko Publishing. Kindle Edition.
[182] McPherson, William. 2015. *Climate, Weather and Ideology.* Amazon.

[183] "Approving the climate security agenda," By Thomas Gaulkin, *Bulletin of the Atomic Scientists,* July 14, 2018
[184] *Scientific American,* June 22, 2018
[185] McPherson, William. 2016. *Sabotaging the Planet,* Amazon
[186] *New York Times,* October 8, 2018
[187] https://www.unenvironment.org/resources/emissions-gap-report-2018
[188] *New York Times,* October 8, 2018
[189] https://www.ourchildrenstrust.org/
[190] http://www.sightline.org/2018/07/09/the-latest-on-the-trial-of-the-century-and-other-courtroom-showdowns/
[191] "The current government have made it very clear they're not interested in doing anything about emissions," according to Matt Drum, the managing director of NDEVR environmental. *Guardian,* September 28, 2018.
[192] "Canada is still quite a distance away from meeting its own commitments under the Paris accord, the latest UN emissions filing shows," *Globe and Mail,* April 18, 2018
[193] A leading candidate for president of Brazil, Jair Bolsonaro, has vowed to withdraw his country, the world's seventh-largest emitter of greenhouse gases, from the pact. "To the extent that we get these narrow-minded, so-called nationalist, populist leaders, we could have a big

problem. Brazil, with its huge area of forests, is going to suffer terribly from climate change," according to ," said John P. Holdren, who served as President Barack Obama's chief science adviser. *New York Times,* October 8, 2018.

[194] UNEP. 2017. *The Emissions Gap Report 2017.* (NDC refers to "Nationally Determined Contributions," the pledges by parties to the Paris Agreement. The U.S. pledge under President Obama was 26-28% reduction by 2025. President Trump, of course, reneged on this commitment.)

[195] *New York Times,* December 7, 2018

[196] *Deutche Welle,* September 27, 2018

[197] CNBC, December 3, 2018. IIEA refers to the Institute of International and European Affairs

[198] *Washington Post,* December 10, 2018

[199] Morton, Timothy. 2013. *Hyperobjects: Philosophy and Ecology after the End of the World.* University of Minnesota Press

[200] "The gospel of climate change: Green pastors bringing environmentalism to evangelicals," Ethan *Sacks*, April 22, 2018, NBC News

[201] *Guardian,* August 27, 2018

[202] Quirk, Joe. 2018. *Seasteading: How Floating Nations Will Restore the Environment, Enrich the Poor, Cure the Sick, and Liberate Humanity from Politicians* (p. 25). Free Press. Kindle

Edition.
[203] Lifton, Robert Jay. 2018. *The Climate Swerve: Reflections on Mind, Hope, and Survival.* The New Press. Kindle Edition.
[204] Joel Wainwright; Geoff Mann. 2018. *Climate Leviathan: A Political Theory of Our Planetary Future.* Verso.
[205] Robinson, Mary. 2018. *Climate Justice: Hope resilience, and the Hope for a Sustainable Future.* Bloomsbury Publishing
[206] *Guardian,* June 14, 2018. Pope Francis did not escape the criticism of collapse ideologues with his reasonable approach. "In promoting rapid decarbonization, Pope Francis has ignored the tremendous human benefits that have been the result of fossil fuel use coupled with human ingenuity. He also, surprisingly, ignores the billion or so people who live in the worst state of poverty because they lack access to commercial energy." *Papal Fallibility on Climate Change,* by William O'Keefe, *Inside Sources*, June 18, 2018. This argument about poverty is found throughout the collapse ideology literature.
[207] *New York Times,* December 3, 2018
[208] Mitchell, Timothy. 2011. *Carbon Democracy: Political Power in the Age of Oil* (p. 253). Verso Books. Kindle Edition.
[209] Andreas Malm. 2016.*Fossil Capital: The Rise of Steam Power and the Roots of Global Warming.* Verso.

[210] Hamilton, Clive. 2018. *Defiant Earth: The Fate of Humans in the Anthropocene* (p. vii). Wiley. Kindle Edition.

[211] "Alien apocalypse: Can any civilization make it through climate change?," *Science News*, June 4, 2018

[212] *New York Times,* August 23, 2018

[213] The IPCC in its special report of October 2018 estimated that the cost of carbon could range from $135 to $5700 a ton depending on the rate of emissions reductions by 2100. See Appendix for full report.

[214] *Vox,* September 27, 2018

[215] *Guardian,* December 9, 2018

[216] *New York Times,* October 8, 2018

[217] CBS *60 Minutes,* October 14, 2018

[218] Ibid.

[219] Rogers, Norman. 2018. *Dumb Energy: A Critique of Wind and Solar Energy* (p. 34). Dumb Energy Publishing. Kindle Edition.

[220] *New York Times,* October 10, 2018

[221] "Rising insurance costs may convince Americans that climate change risks are real," *The Conversation,* October 22, 2018

[222] *Scientific American,* October 17, 2018, adapted from the IPCC Report, see Appendix

[223] *Guardian,* October 24, 2018

[224] "Why Climate Change Skeptics Are Backing Geoengineering," Katherine Ellison, *Science,* 03.28.18

[225] Johansen, Bruce. 2015. *Eco-Hustle! Global Warming, Greenwashing, and Sustainability: Global Warming, Greenwashing, and Sustainability* (p. 111). ABC-CLIO. Kindle Edition.
[226] Hamilton, Clive. 2018. *Defiant Earth: The Fate of Humans in the Anthropocene* (p. 146). Wiley. Kindle Edition.
[227] *New York* Times, June 30, 2018. Two collapse ideologues have cited this research: the Competitive Enterprise Institute and Joseph Bast of the Heartland Institute.
[228] Migration of other species has already begun, and will get more pronounced in the decades to come.
[229] *Guardian,* October 5, 2018
[230] *Community-Led, Human Rights-Based Solutions to Climate-Forced Displacement,* by Amber Moulton, Salote Soqo, and Kevin Ferreira. UUSC, 2018
[231] Ibid.
[232] *New York* Times, June 12, 2018
[233] https://rebellion.earth/
[234] Gray, John. 2002. *Straw Dogs.* Granta Books.
[235] Marshall, George. 2014. *Don't Even Think About It: Why Our Brains Are Wired to Ignore Climate Change* (p. 207). Bloomsbury Publishing. Kindle Edition.
[236] Hamilton, Clive. 2018. *Defiant Earth: The Fate of Humans in the Anthropocene* (p. 117). Wiley. Kindle Edition.

[237] https://www.globalresearch.ca/climate-change-saving-the-planet-saving-ourselves/5624919
[238] Ozarko, Deb. 2018. *Beyond Hope: Letting Go of a World in Collapse* (p. 7). Deb Ozarko Publishing. Kindle Edition.
[239] Davies, Kate. 2013. *The Rise of the U.S. Environmental Health Movement* (p. 5). Rowman & Littlefield Publishers. Kindle Edition.
[240] AP, November 30, 2018
[241] Ozarko, Deb. 2018. *Beyond Hope: Letting Go of a World in Collapse* (p. 7). Deb Ozarko Publishing. Kindle Edition.
[242] Ibid.
[243] Riley E. Dunlap and Robert J. Brulle.2015. *Climate Change and Society: Sociological Perspectives* (pp. 357-358). Oxford University Press. Kindle Edition.
[244] Kolbert, Elizabeth. 2014. *The Sixth Extinction: An Unnatural History* (p. 267). Henry Holt and Co.. Kindle Edition.
[245] Richard Heinberg. 2015. *Blackout.* Kindle Edition.
[246] *Scientific American,* 2009
[247] Angus, Ian. 2018. *Facing the Anthropocene: Fossil Capitalism and the Crisis of the Earth System.* Monthly Review Press. Kindle Edition.
[248] *New Yorker,* November 26, 2018
[249] Harari, Yuval Noah. 2018. *21 Lessons for the 21st Century* (p. 311). Random House

Publishing Group. Kindle Edition.
[250] Spratt, David and Ian Dunlop. 2018. *What Lies Beneath: The Understatement of Existential Climate Risk.* Breakthroughonline.org.au

Made in the USA
San Bernardino, CA
09 January 2019